Building Bridges

Selected papers from the 11th World Environmental Education Congress

Jan Činčera, and David B. Zandvliet, Editors

ISBN 978-1-64504-320-1 (Paperback)
ISBN 978-1-64504-321-8 (Hardback)
ISBN 978-1-64504-322-5 (E-Book)

Library of Congress Control Number: 2024930274

Printed on acid-free paper

11TH WEEC
BUILDING BRIDGES

https://www.diopress.com

Table of Contents

Chapter 1

Building Bridges in Times of Climate Urgency

Jan Činčera, Masaryk[1], and David B. Zandvliet[2]

Since 2003, the World Environmental Education Congress (WEEC) has been a globally important event in the field of Environmental and Sustainability Education. The goal of the WEEC network is to promote the notion that education is key in the promotion of social and ecological sustainability on a global level. In keeping with this idea, all groups and actors (including public, private, nonprofit and for-profit members) are encouraged to participate in the congress, including the academic community, but also non-academic participants. In valuing this openness, different levels of expertise and different approaches are accepted for all aspects of the WEEC events.

The idea for this book follows a tradition that began after the 9th Congress of the WEEC (held in Vancouver, Canada), where it was decided to publish a book of selected papers after the congress. This tradition was also adopted for the 10th WEEC event held in Bangkok, Thailand, and continues its trajectory with this volume (the third) in a series of publications connected to the World Environmental Education Congresses. All submissions accepted for this book were previously accepted for presentation at the 11th WEEC and subjected to a second round of anonymous peer review. All submissions address one or more of the interconnected congress sub-themes, and in the spirit of supporting *praxis*, submissions were directed towards both an educator and/or researcher audience. At a minimum, each chapter describes original research, curriculum development, and/or a case study.

1 University, Faculty of Social Studies, Department of Environmental Studies, Joštova, Czech Republic
2 Professor and UNESCO Chair Simon Fraser University, Vancouver Canada

Building bridges

In their paper "Building bridges, not walls", Braus et al. (2022), concluded:

> "EE is transdisciplinary by nature – a characteristic that has always pushed the field to work broadly. Given the many networking tools in today's connected world, it is exciting to consider how to achieve more by leveraging collective efforts. If we continue to think about EE as an ecosystem, we have the opportunity, and indeed likelihood, to do more, learn more, and accomplish more on the road to creating a more sustainable, equitable future for all." (p. 328)

The concept of "building bridges" between various stakeholders, regions, and groups in the field of environmental and sustainability education resonated with the theme for the 11th World Environmental Education Congress, held in Prague, Czech Republic, 14th–18th March 2022.

The 11th World Environmental Education Congress was organized by a consortium of organizations, Masaryk University, the WEEC Network, and the AIM Group. The Congress targeted the broad scope of the environmental and sustainability education (ESE) community, including scholars, practitioners, and administrators. Its central theme was "Building Bridges". Altogether, 524 scholars, practitioners, and administrators participated in the Congress. The goals were to promote the reflection and development of the ESE field, to provide a learning experience for its participants, to promote cooperation and networking between theory and practice across various discourses and geographic regions, and to provide a friendly and enjoyable experience for the participants.

Despite the recent pandemic and the outbreak of war in Ukraine, which transformed part of the Congress venue into a refugee center, the Czech consortium organized a considerably large and successful event consisting of 15 thematic sessions organized both onsite and online, plenary sessions, field excursions, field sessions, and side events. Jan Činčera, the head of the organizational team, shares his insights from his experience with the conference below:

What did we learn?

The participatory approach is crucial.

It would be paradoxical to speak about the need for participatory learning in ESE and then do the opposite when preparing for our professional meetings. Rather than to organize the congress top-down, we used a participatory, bottom-up approach. Specifically, we invited a high number of scholars from all over the world to "get on board" and then offered them the opportunity to define and lead their thematic sessions. While we offered support and provided basic guidelines, we repeated the "your session is your congress" mantra to allow the thematic chairs enough freedom to use their expertise, experience, and contacts for the best.

Overall, this strategy worked very well. While some problems emerged (mostly due to the COVID situation), many people felt ownership for their part of the congress. Further, this organizational strategy made the thematic sessions smaller, more friendly, and more supportive of networking. Based on this idea, we believe the participatory approach should be the way forward. We believe that large professional meetings (e.g., WEEC or NAAEE) should not belong to any single organization or professional secretariat but to the whole environmental and sustainability education community. As it is, the entire community should get a chance to participate in shaping and structuring meetings.

Building bridges between theory and practice.

To fulfill the central theme, we decided to increase the involvement of ESE practitioners, e.g., environmental educators, scouts, program designers, and administrators. We formed a considerably broad team representing the full spectrum of roles existing in the field. As a result, not only respected scholars but also practitioners got a chance to shape the program and define their part of the congress. Some educators prepared interesting workshops, while others provided various "field sessions"—thematic sessions shaped and managed by the environmental centers in their own facilities or outdoors.

This effort was highly successful. We received positive feedback from the organizers of all field sessions and from participants. The "field sessions" and outdoor excursions were the most often mentioned recommendations for what to keep from the congress for future events in the post-conference survey. We realized there is no need to confine the congress in a conference center, at least not for the entire time. After three days of presentations and indoor workshops, the participants were happy to have a change of environment, to spend time in smaller groups, and to reflect on the field from another perspective. Perhaps we do not need super-formal congresses (in congress centers) as much, as we need to be together in more reflective, non-formal environments. Perhaps future meetings might be 'wilder', as Jickling et al. (2018) and others suggest. Based on this idea, we believe that building bridges between scholars and practitioners may be crucial for fostering better communications in our field. To achieve this, practitioners need to have a part in planning the event to shape it in their own way.

Supporting inter-generational understanding.

While WEEC also featured other workshops, group meetings and events, the most important 'side event' of the WEEC was the 'Youth Congress', organized simultaneously with WEEC for more than 120 young participants. Later in the meeting, this event merged with the larger congress, and students presented their messages to WEEC participants. Youth delegates participated in a joint panel discussion on "building bridges" across the generations. Participants of both congresses highly appreciated this merged event. The involvement of young people became a robust and emotional part of the congress. *"Can you hear us? Really?"* one of the presenting students asked the audience. Perhaps this is what the role of other professional meetings in our field should do: provide opportunities not only to "hear the youth", but to speak and initiate a fruitful collaboration across the generations.

Online experiences do not substitute the need for meeting face-to-face.

Due to the post-COVID situation and an effort to minimize the WEEC's carbon footprint, we organized the event in a hybrid format (online and face-to-face). However, while the onsite participants reflected on the Congress as highly successful, the reflection of those who participated online was relatively modest (but not low). We realized that, regardless of the effort and technological advances, online participation could not yet replace the experience of meeting, chatting with colleagues, or enjoying the sessions in the countryside.

Summary and Overview of Other Chapters

Shaping the congress as a joint project with practitioners, youth, and the entire professional community proved to be the way to make this event meaningful. We believe that the WEEC community will flourish when 'building bridges' and will likewise suffer if we stay in isolated subgroups with narrow discourses, professional orientations, or in specific regions or specializations.

The book you have in your hands provides a potpourri of some of the ideas communicated at the 11th WEEC. At the same time, it could be interpreted as a reflection of its "building bridges" theme: it contains chapters reflecting ESE policy, theory, research, and practice. It may be a wild collection, but we hope you will find it inspiring. Presented briefly here is a summary of the chapters that follow:

In Chapter 2, Ulrike Pisiotis and Deirdre Hodson share research entitled *Bridging the Gap between Research and Policy in Learning for Sustainability*. The authors relate that well-defined policies are key to effective learning for the environment and sustainability. Based on the experience and perspective of policymakers from the European Commission, this chapter explores lessons and identifies success factors and challenges related to good policy design for learning about sustainability. A focus is put on the policy–research nexus.

In Chapter 3, Agnieszka Kozłowska shares a study entitled *Climate Change Education in the National Curriculum Framework: Why Is It So Difficult?* Kozlowska relates that climate change is one of today's most serious challenges, and climate change education is an important tool in combating the climate change crisis. The study aims to decompose the reasons for CCE's absence in national curricula by looking at the case of Poland, where the curriculum framework does not include CCE or includes it to a marginal extent. Eisner's null curriculum concept is used as the pedagogical theory.

In Chapter 4, Edith Planche includes research entitled *Art and Ethnology in the Service of Environmental Education for a Sensitive Anchoring in the Service of Resilience to the Anthropocene.*

Her case study implements environmental education through culture, in partnership with local authorities, schoolchildren or children in their leisure time and adults. Her work reflects this double trajectory with a first anthropological part and a second pedagogical part, including workshops linking theory and practice. The anthropological part makes the hypothesis that the ruptures between Science–Art and Nature–Culture are of the same epistemological strain at the origin of the disconnection between humans and the environment.

In Chapter 5, Éva Neumayer and co-authors share a case study entitled *Biolearn—International Project about Biomimicry in Education for Sustainability.* The authors relate that Biomimicry is a design method that uses knowledge from nature as inspiration for sustainable design. It is an interdisciplinary approach that brings together nature, biology, design, and technology. Similar expressions are bio-inspired, biomimetics, and bionics. The aim of the work was to elaborate educational modules for students aged 11–16 and their teachers and introduce them to biomimicry thinking – a way that can help to make our world more sustainable. The modules were made available in five languages: Czech, English, Dutch, Hungarian and Slovakian.

In Chapter 6, Elena Pagliarino, Bianca La Placa, and Gabriele Ardizio share a work entitled *Civic Participation in Rural Areas during the Pandemic: Reflections on the Community Castles Project.* Their study investigates a community civic initiative (the Community Castles Project) developed in the rural northwest of Italy during the COVID-19 pandemic. The project was built around the concept of physical places (two medieval castles) that can become community centers for cultural, recreational, and decision-making activities. The chapter focuses on the narrative of the project, and its main objective was to identify which actors and processes are relevant to fueling participation in rural areas.

In Chapter 7, Elvio Kamiyama and co-authors present a study entitled *Young (K–5) Environmental Agents Course: The Importance of Remote Practical Elementary School Environmental Education.* The case study describes an early childhood environmental education course that was administered online to a rural public school in the state of Rio de Janeiro, Brazil. The aim of the course was to teach the principles of biomes through practical and online activities. The course consisted of carrying out several practical environmental education activities over a period of 3 months that students carried out in their homes and in their immediate natural surroundings in order to create greater awareness and sensitization to core themes. The syllabus included theoretical, explanatory online classes complemented by streaming videos, theoretical and practical activities using delivered course materials, and common materials the children had in their homes and collected in nature.

In Chapter 8, Andreas Brunhold shares research entitled *Assessment and Management of Flood Risks in the EU - Goals and Limitations Depicted at the German River Murr*. Brunold relates that floods in parts of Germany in July 2021 are clear indicators that the ongoing anthropogenic climate crisis leads to changes in the hydrological cycle. His case study concerning the Murr, a small German river with high flood risk, argues that the flood risk management of the EU and its member states has further to be constituted as a key environmental challenge of the 21st century. The case study builds on data collected using a standardized questionnaire. The survey was conducted as a reaction to the application of technical flood prevention measures in different villages and towns in the Murr area. The questionnaire was addressed to different kinds of experts in the field of flood prevention, such as water districts, lower water authorities and citizens' groups, among others.

In Chapter 9, Valerie Lewis shares her work entitled *The Take 1 Programme: A Case Study Approach to Embedding Education for Sustainable Development in Learning and Teaching*. Her work highlights the importance of education for sustainable development (ESD) and emphasizes the need for its inclusion in the formal and non-formal curriculum. The chapter provides an overview of the design and application of the Take 1 Programme as a mechanism for embedding ESD in learning and teaching in Irish second-level schools. It demonstrates an approach to engaging with ESD through the Sustainable Development Goals using a methodology mapped to the learning outcomes of curriculum subjects.

In Chapter 10, Hideaki Kurishima and co-authors share a study entitled *Development and Trial of a Climate Change Education Program Featuring a "Carbon-Neutral Simulator" for Schools*. The study relates that in order to achieve carbon-neutral regions, it is crucial for younger generations to take responsibility at the regional level and comprehend the pathway toward decarbonization while considering appropriate countermeasures. Therefore, it is necessary to design climate change education (CCE) programs for children. The study developed a school version of a "carbon-neutral simulator". that explored carbon-neutral strategies at the regional level. Through this education program, students were readily able to grasp climate change and decarbonization concepts.

In the final chapter, Kateřina Jančaříková and co-authors share a study entitled *Forest Kindergartens in the Czech Republic.* Their study outlines the movement in support of forest kindergartens in the Czech Republic during the last 20 years. The chapter describes the incredible progress of the movement (initiated by Emilie Strejčková in 2003), relating that there are now about 200 of them in the Czech Republic, attended by over 3300 children. The chapter is based not only on the usual scientific approaches but also on the life experiences of the co-authors, respectively a teacher, director and university lecturer, all linked to one of the first forest kindergartens in the Czech Republic.

We wish you an enjoyable and inspiring read.

References

Braus, J. A., Heimlich, J. E., Ardoin, N. M. & Clark, C. R. (2022) Building bridges, not walls: exploring the environmental education ecosystem. *Applied Environmental Education & Communication, 21*, 4, 319-330. DOI: 10.1080/1533015X.2022.2115226

Jickling, B., Blenkinsop, S., Timmerman, N., & Sitka Sage, M. (Eds.). (2018). *Wild pedagogies: Touchstones for re-negotiating education and the environment in the Anthropocene.* Palgrave Macmillan.

Chapter 2

Bridging the gap between research and policy in learning for sustainability

An example from EU policy

Ulrike Pisiotis

Introduction

Learning for sustainability[1] is an educational agenda that 'naturally' transcends the sphere of education and training and touches on other policy fields, most notably environmental politics. This is evident by the fact that in many countries, ministries and other government agencies that do not primarily deal with education nevertheless create and engage in educational and outreach initiatives relating to the environment and sustainability. This complexity of responsibilities alone—before even considering the inherent complexity of sustainability as such—often requires additional efforts by policymakers to design and negotiate well-defined and coherent policies on learning for sustainability compared to other agendas in education and training.

When it comes to the expected outcomes of learning for sustainability, the question to policymakers is whether learning for sustainability policies primarily benefits the environment and sustainability, the overall quality of education, or both of these agendas. While research points to clear advantages of seeking out synergies over an either/or attitude, in practice, those creating learning for sustainability agendas do not always possess the knowledge and tools necessary for educational change (Mulà & Tilbury, 2023) understood as a transformation that reaches the micro-, meso- and macro-level of education—in short, the whole system. In addition, even if learning about

1 Apart from specific references, the term "learning for sustainability" is used throughout the paper standing in for closely related terms used in the field such as environmental education, environmental and sustainability education (ESE), education for sustainable development (ESD), and climate education.

sustainability is under the responsibility of one government agency, it is often one among many policy agendas to be dealt with by that agency or even the same person.

In the case of the European Commission, there is an additional layer of complexity that policymakers in the field of education and training face: policy is not binding but recommended since responsibility for education and training policy lies with the EU Member States. In practice, this has several effects:

- Commission policymakers engage with an especially wide stakeholder constituency: policymakers from (27) Member States, European Economic Area countries (as most legislative proposals are applied in the EEA area), social partners, and non-governmental organisations; all representing different education systems with different starting points and evolution;
- Legislative initiatives are compromises of all EU Member States, therefore increasing take-up in the individual countries but necessarily altering—and in some instances what researchers call 'watering down' (Stratford & Wals, 2020)—original legislative proposals and research evidence;
- Policy borrowing and uptake of EU education priorities play an increasing role in national policymaking, as is the case with other international networks and players in education (such as OECD);
- Expectations on the impact of EU recommendations on national and regional policy and practice arising from the institutional clout of the EU are primarily met with formats of collaboration, funding initiatives, and other voluntary support actions from the EU side;
- Impact monitoring is further removed from practice on the ground than at the national or regional level (and arguably harder to translate into policy adaptation).

The different responsibilities and expectations with regard to learning for sustainability policies described so far underline two actualities for policymaking: the difficulty of linking the complexity of sustainability to that of the education system in order to create viable synergies; and the multi-faceted or "messy" (Rickinson & McKenzie, 2021b, p. 481) process that is policymaking. Given the 'wicked' problems the world is facing—among them climate change and environmental degradation—it is not surprising that some even consider policymaking "a formidable task" (European Commission, 2002e, p. 2).

The challenge for policymakers is getting from these premises to well-informed, joined-up policies in education and beyond. Ideally, education is part of *all* policy documents relating to sustainability, including those beyond education, such as transport, industry, research policies, etc. (Glavic, 2020). Within the policy field of education, it is the kind of policymaking that is able to incorporate and weave together the complexities of both sustainability and wider educational change and address the multiple entry points for sustainability into education and their relation and interdependence (e.g., curriculum change, teacher education, and assessment) that is more likely to have a lasting impact on embedding sustainability into education (European Commission, 2022b). While selected space- and time-bound measures have their merit, there is general agreement that comprehensive strategies that address several—often connected—elements in education and training are necessary for systemic change (UNESCO, 2019). Hence, what applies to education policy in general is also true for learning for sustainability: "It is not sufficient to simply identify a problem or challenge that needs to be tackled and select an education policy solution" (OECD, 2020, p. 4). Instead, a "compelling and inspirational vision" that "can engage stakeholders around it" is needed (OECD, 2020, p.4). Achieving a sustainable world that respects the boundaries of our planet is a very compelling and inspirational vision. Yet, embedding sustainability in education and training is still far from being systemic (European Commission, 2022b) because linking concerns for sustainability and improving education quality is so far underdeveloped compared to treating sustainability as an add-on to education programmes.

Research forms a substantial part of the evidence that policymakers use to arrive at well-informed policies. Research use has been increasing in importance since the 1990s with a global trend for evidence-based or evidence-informed policymaking (Rickinson & McKenzie, 2021a). However, there is rarely a clear-cut and linear development from research input to policy output. Firstly, research itself is a diverse phenomenon that contributes to policymaking in different forms. Secondly, policymakers rely on sources other than research for input and evidence in their work, as well as to account for the political process that constitutes policymaking (European Commission, 2022g; Rickinson & McKenzie, 2021b).

Joining a growing number of publications on the research–policy or evidence–policy nexus, this paper explores lessons and identifies success factors and challenges related to learning for sustainability policy design based on the experience and perspective of policymakers[2] from the European Commission. It starts with shedding light on the process leading up to a major EU policy initiative in the field. Thus, it responds to calls to go beyond investigating policy texts and looking into policy processes (Rickinson & McKenzie, 2021b). Describing this particular policymaking process it can serve to further illustrate "policymakers' perspectives on evidence use" (Shaxson & Boaz, 2021, p. 521) that can contribute to advancing learning for sustainability. It then identifies evidence gaps from the EU policymaker's perspective. Finally, it discusses possible formats of policy learning as well as research–policy cooperation in the field of learning for sustainability.

The making of recent major EU policy initiatives in sustainability education

In June 2022, EU Member States adopted a Recommendation (i.e., non-binding EU legislative initiative) on learning for the green transition and sustainable development based on a proposal by the EU Commission for such a recommendation from January 2022. The EU Commission also published GreenComp, a framework for sustainability competences which can be used in education and training programmes and policies in formal, non-formal and informal settings. Both initiatives relied on substantive and diverse input and evidence, including from research.

Since the early 2000s, the European Commission has been implementing its 'better regulation' approach across all policymaking (European Commission, 2022f). Impact assessments and evaluations, as well as stakeholder consultations, form consistent parts of the policy-making process. The *better regulation toolbox*[3] sets out the steps needed for designing policies that consider the opinions and positions of stakeholders and interested parties, including organisations and individuals.

The preparation of the EU Commission's proposal for a policy recommendation on learning for sustainability illustrates how the better

[2] The author and Deirdre Hodson.

[3] https://commission.europa.eu/law/law-making-process/planning-and-proposing-law/better-regulation/better-regulation-guidelines-and-toolbox/better-regulation-toolbox_en

regulation approach plays out in practice. The initiation of the policy is clearly linked to the European Green Deal, the EU's strategy for becoming climate-neutral by 2050. Adopted in 2019, the Green Deal has, by virtue of its foundational and guiding character-driven policy development across a range of areas related to a green transition, education included. In the Green Deal communication, the role of schools and other education institutions "to engage with pupils, parents, and the wider community on the changes needed for a successful transition" was put into focus (European Commission, 2019, p. 19). Following up on that assertion, two subsequent EU policy documents, the biodiversity strategy and the European Education Area strategy, announced the Commission's intention to propose a recommendation on 'education for environmental sustainability' to the Council of the EU, i.e., the heads of governments of the 27 EU Member States. The endorsement of a green dimension of education in the EU's overarching multi-policy strategy for sustainability (the Green Deal) and the EU's comprehensive, multi-area strategy on education (the European Education Area) were strong anchors and drivers for our specific policy proposal. While recognising Member States' efforts in the area of learning for sustainability[4], including the existence of policy documents at different levels, the proposed EU Recommendation was meant to bring political momentum to the sustainability education agenda and to renew and strengthen efforts by EU countries in that area.

It is worth pointing out the challenges that we faced in the development of the policy proposal, in addition to the organisation-specific challenges mentioned in the introduction. They included:

- Learning for sustainability being a new priority for the Commission there was no organisational legacy to build on, such as previous expert groups or positions;
- Internationally as well as in many cases nationally, learning for sustainability had been an established field already for decades, starting with environmental education and being shaped in particular by UNESCO and its work on education for sustainable development (ESD);
- As policymakers in the area of school education, we developed a life-long learning initiative, thus adding complexity of internal coordination, input of evidence and use of terminology.

[4] In the text: "policies and programmes of learning to support the green transition and to promote learning for sustainable development" (Council of the European Union, p. 2).

The Commission's ideas and objectives for preparing a legislative proposal in the area of learning for sustainability were first outlined to the public in April 2021 on the Commission's *Have Your Say* portal,[5] where individuals, organisations and businesses can share their views on new EU policies and existing laws. The announcement was complemented by a roadmap[6] that stated the purpose, scope and limitations of the proposal. The roadmap first specified the political and policy context of the planned recommendation, defined the problem that the proposal would address and elaborated on the EU-added value for this proposal to ensure it is in line with the principles of subsidiarity and proportionality laid down in the EU treaties. The roadmap then stated the aim of the initiative and details how information would be gathered to prepare it.

As is common in legislative proposals, the public had four weeks to react to the Commission's announcement. On the roadmap for the recommendation, we received 31 statements by organisations and individuals, the majority originating in non-governmental organisations active in areas including environment, education, youth, and social integration. The following visual of feedback statistics related to the

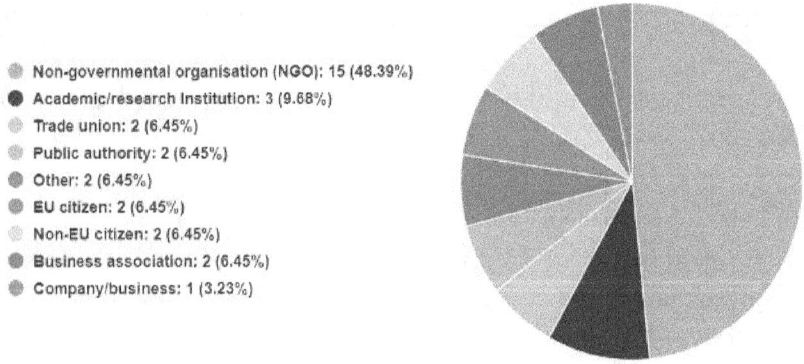

Figure 1: Proportion of feedback statements to the roadmap by category of the respondents. Source: https://ec.europa.eu/info/law/better-regulation/have-your-say/initiatives/12985-Environmental-sustainability-education-and-training/feedback_en?p_id=23832124

5 https://ec.europa.eu/info/law/better-regulation/have-your-say_en

6 https://ec.europa.eu/info/law/better-regulation/have-your-say/initiatives/12985-Environmental-sustainability-education-and-training_en

roadmap can be found on the *Have Your Say* portal:

The feedback statements covered a range or continuum of questions from and beyond directly education-related issues. Among others, the importance of non-formal education, curricula development, and life-long learning was pointed out, as were the food system, the need for healthy (school) meals, and product labelling. This feedback was then taken into account for the next major step in public consultation on the Commission's proposal: a public survey consisting of 10 questions relating to learning for sustainability and available in all official EU languages. The survey ran from June to September 2021 and received 1368 (valid) replies from individuals and organisations, with the majority coming from EU citizens (46%), followed by academic and research institutions (21%). Associations and (non-profit) organisations that often collaborate with academic research constituted 8% of respondents.

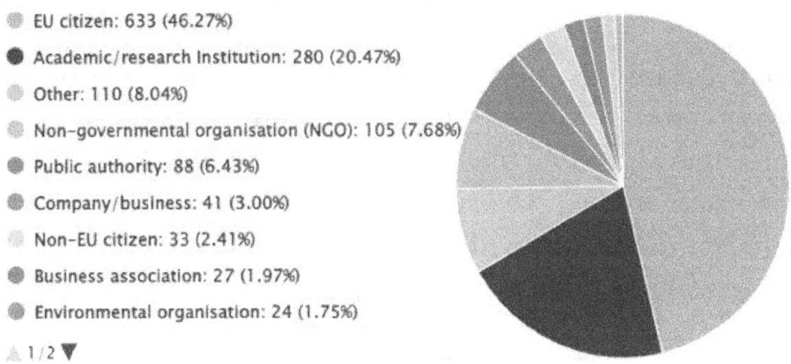

Figure 2: Proportion of public survey respondents by category. Source: https://ec.europa.eu/info/law/better-regulation/have-your-say/initiatives/12985-Environmental-sustainability-education-and-training/public-consultation_en

The replies to the survey were analysed following a thorough quality assurance of the data (including checking for identical entries and cleaning the data[7]). Following translation, open-text answers were coded manually (based on a set of identified codes/categories) before running a frequency analysis to concisely complement responses to the multiple-choice questions. The results of the public survey

[7] For instance, if participants provided more than one response to single-response questions, these cases were considered invalid.

were documented in the Staff Working Document accompanying the Commission's proposal for a Council recommendation (European Commission, 2022e).

To gather more direct input from stakeholders, we also held a series of consultation workshops designed around key themes for the proposal on learning for sustainability, such as effective policy levers and measuring impact, collaboration, and partnerships, as well as teacher education. These workshops were attended by a broad variety of stakeholders from academe, education and environmental organisations, international bodies (UNESCO, the Council of Europe), teachers and social partners such as teacher trade union organisations. The participative format of the workshops, with their group discussions and online collaborative tools open to all participants, allowed us to capture a broad spectrum of views on the topic of sustainability education and the Commission's proposal in particular.

The role of research

In a recent communication, the EU Commission described scientific evidence as "vital to establishing an accurate description of the problem, a real understanding of causality and therefore intervention logic; and to evaluate impact" (European Commission, 2021, p.3). The better regulation toolbox provides for science and research as an integral part of preparing policy initiatives. Before describing the use of research and evidence (beyond the data gathered in the public survey and consultation workshops) for preparing the Commission's proposal on sustainability education, it is worthwhile to consider the different forms in which research enters the policymaking process.

It is not surprising that policymakers "with higher levels of education or some experience of research" (Lysgaard et al., 2016, p. 326) are more likely to use academic evidence in and for their work. At the same time, policymakers, even if academically inclined, operate in a complex environment where they address political, organisational, outreach, and communication demands—often in a time-sensitive manner. Research might and often is contributing to those activities. However, "even the most compelling and relevant research findings may fail to penetrate the policymaking process" (Orland, 2009, p.115). From the limited perspective of the EU education policymaker, it seems that Lysgaard et al. (2016) bring into focus compelling arguments with regard to the nature of research that is more likely to be considered by policymakers. This happens when research is of "high quality" and "from a credible source"; "provides clear and uncontested findings"; "has been commissioned" or otherwise enjoys "high-level political support"; responds to "local priorities, needs and contexts"; "is timely and relevant to policymakers"; is "concise, jargon-free, and visually appealing" (p. 326). Timeliness and relevance are often synonymous with fitting into the political and policy narrative (Rickinson & al., 2019). The phenomenon could also be described as research being "aligned with current ideology" (Lysgaard et al., p. 326). More neutrally, it is referred to as (policy) background and context, including in the case of the Commission proposal on sustainability education, and it is taken up in the roadmap and the proposal text itself. Furthermore, direct contact between researchers and policymakers, individuals or agencies acting as knowledge brokers is likely to increase the uptake of research (Lysgaard et al., 2016).

The intention of establishing closer links between research and policymaking is an affirmative one for both sides: for policymaking, it means building trust in and acceptance of policies, whereas research might be attributed with increased relevance, resulting in a broader and deeper impact. However, there are also risks associated with closer relations between science and policy, such as the possibility of compromising research integrity for policy expectations. Instrumental and simplistic ways of conceiving of research as driving policy or politics shaping research and knowledge in one-dimensional ways are alarming when they constrict academic freedom, usually playing out as support or the lack of it through directional public funding (Boswell & Smith, 2017). Tying research funding to 'useful' outcomes and expected impact on policy and society might limit researchers' potential "to promote ideas that are out-of-line with government policies" (Boswell & Smith, 2017, p. 5). Taking this a step further, we would arrive at the possibility of politicised research. While there is no absolute safeguard against misdirected interpretation and use of the research–policy relationship, I would argue that the understanding on both sides (research and policy) of their respective roles and interactions has grown in complexity. This has resulted in multifaceted expectations and interactions which do not eliminate but reduce the risk of impinging on research freedom and communication through policy direction.

Many of the forms in which research becomes part of the evidence used for policymaking are accounted for in the process, leading to the Commission's proposal on learning for sustainability. In addition to inviting feedback on the roadmap and public survey, including from the research community and engaging with researchers in the different consultation workshops, we commissioned several studies and reports on specific topics to help prepare the Commission proposal, thus initiating research *for* policy (Rickinson & McKenzie, 2021b).[8] One of those studies undertook a mapping of policy and practice approaches to learning for environmental sustainability across EU Member States (European Commission, 2022b). Our policy proposal was partly shaped by the study's findings (e.g., the fact that learning for sustainability is not yet a systemic feature in many education and training systems, sustainability competences are rarely defined, good practices often have an expiration date as they are tied to projects and not scaled up). As the pandemic had both a big influence on the way, children experienced education and a direct link to environmental and sustainability concerns, another study analysed the impact of COVID-19 on Education for Sustainable Development (ESD) in the context of the green and digital transition (European Commission, 2022d). Again, the findings, especially those complementing the mapping study (on learning environments, student participation, and partnerships) contributed to inform our policy proposal. Finally, as the EU's Erasmus+ programme funds education projects in all areas, we commissioned a report collecting data on Erasmus+ projects which had established good practices with regard to sustainability education (European Commission, 2022a). Insights from the specific projects, including success and failure in the (frequently multiannual) implementation, enriched our work with accounts of field experience that can otherwise be difficult to obtain.

In addition to the commissioned studies and reports, the position papers, survey results and the thematic discussions in our consultation workshops, we engaged as policymakers in thorough own-initiative reading, studying and training to help us evaluate and weigh the complex input. Research articles on various topics, grey literature and

[8] While not limited to those, there are several research networks that Commission policymakers in education can tap into: (i) the Commission's Joint Research Centre (JRC); (ii) Eurydice, a research-based network of the Commission to gather data from Member States; (iii) the Network of Experts working on the Social dimension of Education and Training (NESET); and (iv) the European Expert Network on Economics of Education (EENEE).

training offers (in-person, webinars and MOOCs) form a continuous feature of our work. For the concrete policy proposal, many of those sources can be found in the accompanying so-called Staff Working Document, which provides background, examples and context to the proposal (European Commission, 2022e). At the same time, while our own-initiative learning as policymakers represents a substantive part of research used for preparing the proposal, it is almost impossible to disentangle in retrospect the exact impact of distinct research items. Internal dynamics are at play as well, with managers or other departments bringing in their specific knowledge and positions. This aligns with views of the process of policymaking as "cyclical, multi-directional, or messy" rather than linear (Rickinson & Mc Kenzie, 2021b, p. 481). It also supports observations by researchers on the long-term effect of research on policymakers' knowledge and attitudes. Lingard (2021), citing Weiss, speaks of "an enlightenment or 'percolation' impact" and argues that research has a lasting impact on the "assumptive worlds of policymakers" (p. 505).

The account provided so far demonstrates that research findings were strongly present throughout the whole policymaking process. Positions papers submitted by researchers or academic organisations and especially discussions in workshops were channels to research evidence, positions, and developments from the field of learning for sustainability. Underscoring the fast and direct impact of research *for* policy (Lingard, 2021), this (direct) exchange also helped to draw our attention to certain agendas as meriting more focus in sustainability education policymaking than so far received. Those included early childhood education and care, non-formal learning, youth engagement, specific competences (such as systems thinking and futures thinking), as well as whole-school or whole-institution approaches. While we had picked up on those areas through our preparatory reading, it is safe to say that they became much more prominent in the subsequent proposal because of the survey and the discussions in the consultation workshops.

The role and strong presence of research in the Commission's work so far on learning for sustainability is equally or arguably even more apparent in GreenComp, the European sustainability competence framework (Bianchi, et al., 2022), which was published along with the Commission's proposal for a Council Recommendation in January 2022. Unlike the Council Recommendation, which is a political statement, the framework is a resource offered by the Commission to education stakeholders for voluntary use and does not require political negotiations. Based on leading research findings in the field of sustainability competences, it was developed over one year under the lead of the Commission's Joint Research Centre (JRC) and is the result of a consensus of "a diverse group of approximately 75 experts and stakeholders" from Europe and beyond, including "experts on sustainability education and lifelong learning from academia and research institutions, youth representatives, educators, policy representatives from EU Member States and NGOs" (Bianchi et al., 2022, p. 7). To arrive at the final version of GreenComp, scoping studies, including a comprehensive literature review on ESD competences, were carried out. They were followed by a series of workshops with the experts and stakeholders to consolidate, refine and validate the emerging framework (Bianchi et al., 2022). As GreenComp is an educational resource with strong foundations in sustainability and sustainability education research, the research impact on its creation is very direct, and the development process is more linear (Rickinson & McKenzie, 2021b) than for the Council Recommendation. GreenComp, developed through the cooperation of researchers and policymakers, is indeed a case at hand for describing the role of research not in a binary fashion between research *for* (i.e., commissioned research) and research *of* policy (i.e., research for its own sake or basic research) but rather as a continuum (Rickinson & McKenzie, 2021b).

At the same time, GreenComp is an excellent example of policy measures being aligned with the current political narrative to be effective, hence the importance of timeliness and relevance of the research that supports those measures (Lysgaard et al., 2016). GreenComp is part of an ecosystem of competence frameworks the Commission has published since 2013 and marks an intentional shift towards sustainability education and competences following the ambitions announced in the European Green Deal. Previous competence frameworks, including digital competences (DigComp), entrepreneurial competences (EntreComp), and personal, social, and learning to learn competences (LifeComp), are all offshoots of the EU policy on key competences for lifelong learning. First adopted by the European Parliament and the Council in 2006, the Recommendation for Key Competences for Lifelong Learning is credited as a catalyst for the shift towards competence-based school education across Europe (European Commission, 2022c). In 2018, i.e., one year before the European Green Deal, an adapted and revised version of the Recommendation was adopted by EU Member States, maintaining the number of key competences (eight)[9] and, by and large, their scope while aligning certain aspects better with the needs of the time. This is why 'communication in the mother tongue' became 'literacy competence', 'mathematical competence and *basic* competences in science and technology'[10] became 'mathematical competence and competence in science, technology and engineering' or 'learning to learn' was upgraded to 'personal, social and learning to learn competence'. However, sustainability was not referenced at the time as a separate competence but rather a cross-cutting concern that manifested in STEM, citizenship, and entrepreneurship competences. Thus, while duly acknowledged, the policy message communicated at the time about the urgency of learning for sustainability was rather subdued and secondary in relation to other competences. The collective shift by decision-makers to embed sustainability in *all* policies following the announcement of the European Green Deal in 2019 provided the necessary political momentum to develop effective policies and guidance on learning for sustainability, including GreenComp.

Evidence-informed policymaking

Notwithstanding the critical role of research and other evidence in policy work, the outcomes of the policymaking process, in particular the policy proposal for a Council recommendation (European Commission, 2022g) and the final recommendation, adopted by the Council (i.e., the EU institution representing EU Member States), is much more than a mere compilation of the diverse input that we solicited. As policymakers, our role is to extract and construct from the rich material available to us what is novel and relevant for our proposal while also linking to existing opportunities (e.g., funding) both at EU and national levels as well as existing policy lines in education and beyond (the role of teachers, the Green Deal). Inside the Commis-

[9] Literacy competence; Multilingual competence; Mathematical competence and competence in science, technology and engineering; Digital competence; Personal, social and learning to learn competence; Citizenship competence; Entrepreneurship competence; Cultural awareness and expression competence.

[10] Italics U.P.

sion, this involved an intra- as well as cross-departmental consultation process to solicit input from other education and non-education policy areas that are relevant for learning for sustainability, including employment, environment, and climate. Assessing and weighing education, sustainability and other arguments, it was at times challenging to work with the perspectives of (non-education) departments. For example, while education policymakers are usually focused on systemic education reforms, other departments may sometimes seize such initiatives as opportunities for making their activities better known to the broader public.

The very process of policy development also places limits on the use of science and research. The reasons are twofold. First, given the variety of demands and tasks policymakers need to address, the attention they can give to research findings is necessarily limited. Second, policymaking is ultimately a political process into which other considerations than purely scientific ones enter. It is now customary to speak of evidence-*informed* policy rather than evidence-*based* policy "to acknowledge the limits of science" (European Commission, 2022g) and research in the complex input that shape new policies.

The influence and limits of research in policymaking are present throughout the different steps of EU policymaking. As described above, developing and drafting the actual policy proposal relies on research input at different stages and in different forms. At the same time, internal discussions and reflection shape the policy proposal that is adopted in the end by the Commission (i.e., the College of Commissioners). The next and final step for the proposal to turn into EU policy involves negotiations and agreement by the Council. As is the case of other policy and legislative initiatives, negotiating the Commission's proposal for a recommendation on learning for sustainability in the Council translated into finding a compromise (text) that worked for the nations present in the EU regarding the organisation of education and training, as well as the evolution of learning for sustainability. The resulting Recommendation adopted by all EU Member States (Council of the European Union, 2022), therefore, displays several changes compared to the Commission's original proposal (European Commission, 2022f). This, however, does not mean that language or proposed measures have necessarily been watered down (Stratford & Wals, 2020). On the contrary, some points were brought into sharper focus, such as the autonomy of teachers to design learning and choose pedagogical approaches or the role of youth work and non-formal education.

The way forward—research and policymaking in sustainability education

Bearing in mind the necessarily limited view of policymaking to which the author of this article has access, it might be helpful to share some observations regarding links and gaps between research and policymaking with a view to strengthening sustainability education policy in particular. Filling those gaps can impact us as policymakers and research funders as it could (i) contribute to developing better and more targeted policies at the EU level and shaping better policies in Member States and (ii) help generate more research directly (e.g., through commissioned studies) or indirectly (e.g., through calls for research such as Horizon Europe).

Critical policy research

The default notion of research and policy as two distinct realms does seem to obscure at times the intrinsic and reciprocal relationship they do nonetheless share. It is customary for policymakers to make use of research in education—with all the caveats described above. However, there seems to be a gap in research when it comes to consistently reflecting policy influence, i.e., the impact of concrete policies on the phenomena and data that is of the most interest to educational researchers. Most phenomena pertaining to learners, teachers, and outcomes are the result of a conglomeration of contributing factors, among them concrete policy interventions—not necessarily those relating only to education. Analysing this relationship more consistently, especially in the field of learning for sustainability, could contribute to improving policymakers' understanding of their work and impact and, hence, improve policies. Along the same lines, Aikens et al. (2016) suggested that more attention be paid to researching 'critical policy theory and methodology', which goes beyond (short-term) impact assessment and instead focuses on how policy measures are "championed, borne out, resisted, or subverted in practice" (Aikens & al., 2016, p. 350).

The impact of place and context

In their review of trends and methods in environmental and sustainability education comprising 215 research articles representing 71 countries, Aikens et al. (2016) also underlined the contextuality of policy development and enactment. With a birds-eye view of sustainability education policies and measures across the EU, we have seen diverse approaches to embedding sustainability into education and training. A better understanding of the different challenges and successful approaches between, but also within, countries is especially pertinent to sustainability education with its emphasis on localisation and community involvement. In our preparation for the Council Recommendation, we have seen various examples of projects and initiatives in different contexts, but it is not always clear why something works in one place and not in another. So, it would be helpful to see such a comparative evaluation.

Improving the quality of education through sustainability

As laid out in the introduction, the scale, breadth, and contestation of learning for sustainability education has led to discussions about its content as much as about its purpose. The question is whether it should serve the environment, the climate and sustainability more generally or whether it should serve to improve educational offers (Rickinson & McKenzie, 2021b; Aikens & al, 2016), including through novel subject matters, learning environments, and actors. Our position as policymakers is to seek out synergies rather than focus on different purposes that will be very hard to justify in reality. Ideally, learning for sustainability improves the quality of education and, consequently, leads to attitudes and actions that improve our planet's health.

In practice, however, it seems more difficult than one would think to combine both purposes, as those designing learning for sustainability education are not necessarily well-versed in the ins and outs of educational change (Mulà & Tilbury, 2023). In research, some attention has been paid to how learning for sustainability improves overall learning ability (Dixon et al., 2017). However, more evidence is needed on how it can support other educational agendas. This 'spillover' effect is, to some extent, already visible in a growing interest in the links between sustainability and well-being. However, more often than not, learning for sustainability is perceived as competing with other agendas rather than supporting them (Aikens & al., p. 348). This is not the fault of sustainability education but rather the design of curricula and education systems that prioritise other agendas. Therefore, widening the evidence base on the synergies between sustainability education and the overall learning experience and considering barriers and enablers to both policy development and enactment will strengthen the case of sustainability education and improve the quality of education overall.

Researchers and policymakers—a relationship in transition?

The relationship between researchers and policymakers in education has been characterised by a range from belonging to separate orbits to drawing attention to the overlaps between the two. Regarding the latter, the most prominent are double functions of some researchers and policymakers as knowledge brokers or consultants or cross-overs between academe and policy (Rickinson & McKenzie, 2021a; Lingard, 2021). On one hand, the descriptions of both separation and overlap between research and policy operate on the assumption of two different cultures. On the other hand, Ginsburg and Gorostiaga point out the heterogeneity within both groups in relation to the actors and work processes that cover a spectrum, from objective and technical to more context- and ideology-driven practices (Ginsburg & Gorostiaga, 2001).

In the field of sustainability education, in particular, there are also calls for a more radical role of researchers than that of a neutral producer and custodian of knowledge and understanding. In accordance with clear aims and urgency ascribed to the field to initiate change beyond education, some have argued to step out of what is perceived as a passive role vis-a-vis policymaking and assume an activist stance (Lingard, 2021). From our perspective as policymakers and with the still fresh experience of developing a proposal for a Council Recommendation, it is noticeable that a transition from perceiving researchers merely as experts to regarding them as critical friends is not only desirable (Mulà & Tilbury, 2023) but has already happened to a large extent. The strong role of research and researchers throughout the proposal's preparation is proof of that. While some might argue that commissioned research, in particular, might be missing critical distance to policymaking, its findings are usually based on non-commissioned research and well-anchored in the research field overall.

Still, we have to concede it might be difficult for the policymaker to judge how critical or friendly a researcher may be in their role as a knowledge broker or adviser. Exposure to diverse researchers and research positions is therefore important for policymakers to develop this kind of judgement. For a global issue, this includes exposure to more perspectives from geographical areas that are underrepresented in research (South America, Eastern Europe, North and West Asia, and North Africa, according to Aikens & al., 2016). There is likely no more effective way than through participation in research congresses to offer awareness expansion through organised formats as well as informal exchange. In our case, this is why participating, for example, at the World Environmental Education Congress (WEEC) was an important learning experience. It offered a good overview of research trends in the field and made us question some of our assumptions.

At the same time, we observed that very few policymakers were present compared to researchers. While researchers frequently called for more exchange with policymakers (judging by the lectures and workshops we were able to attend), the overall low numbers of policymakers at the congress indicate that there is room for improvement when it comes to building awareness and participation. It is indeed such events—which might well have an overarching theme (as WEEC did) but do not follow a specific policy agenda (e.g., summits)—that provide an ideal and necessary context for unfiltered exchange.

The WEEC (which, together with on-site visits, lasted a good week) was already similar to formats likely to improve understanding of the specific challenges and opportunities of each side. Such formats include sustained (over an extended period of time) training opportunities specifically for policymakers and based on their needs (e.g., summer academy). They could also comprise staff exchanges between researchers and policymakers to increase awareness of research and capability to utilize it among policymakers and willingness and capability to engage with policy among researchers. According to the JRC survey, "62% of respondents do not think that policymakers in Europe have the skills to understand and critically appraise scientific evidence, and 63% of respondents do not think that scientific knowledge is synthesised, translated, and formatted in a way in the ecosystem that policymakers can use it easily" (European Commission, 2022g, p. 26).

Another example of the changing relationship between researchers and policymakers is knowledge hubs—often in the form of online platforms—that bring both sides together on a specific topic or even a range of topics and identified challenges. The Commission's *Competence Centre on Foresight*[11] is a case in hand aiming to reinforce evidence use and embed foresight into EU policymaking. Established in 2018, it brings together policymakers across different areas and researchers from within and outside the Commission to generate anticipatory knowledge related to various thematic fields and policies. The specific format allows for own-initiative research (e.g.,"picking up weak signals of change and events that might have significant future implications for the EU"[12]) as well as analysis based on the needs of specific Commission departments, representing distinct EU policies. The latter can be arguably construed as an impediment to independent research; on the other hand, as argued above, commissioned research usually draws its insights from the overall (including non-commissioned) research field. A more open format is *Futures4Europe*, an online platform and "co-working space" for "futurists, foresight experts, EU policymakers and citizens"[13]. Supported by the Commission, *all* participants are invited to share their perspectives on Europe's future, thus not only bridging the research policy gap but also including voices from other domains, such as art and the public at large. While perhaps not yet representing "collective research and praxis" where both researchers and policymakers expand "their traditionally distinct roles" (Ginsburg and Gorostiaga, 2001, p. 191), it definitely facilitates dialogue between both groups and through the inclusion of other actors also limits the risk of "overblown deference" between researchers and policymakers (ibid, p. 195) which prevents true exchange.

Supporting different formats of exchange between policymakers and researchers, ranging from ad-hoc to longer-term engagements, will likely contribute to decreasing this kind of alienation. From our side, we continue to work with research and researchers for the implementation of the recommendation with the aim to better link academe with the policymaking sphere. We run a peer learning group for policymakers from Europe, where we regularly invite researchers

[11] https://knowledge4policy.ec.europa.eu/foresight_en

[12] https://knowledge4policy.ec.europa.eu/foresight/about_en, accessed on 24/06/2023.

[13] https://www.futures4europe.eu/, accessed on 22/06/2023.

as guest speakers. The group has been supported by two researchers in the field (Daniella Tilbury and Conor Galvin), who provide input papers that contribute to key messages and other deliverables that the group produces and which are publicly available[14] and of potential interest to researchers.

Conclusion

Both research and policymaking contribute to the creation of prevalent narratives and new paradigms. While the contribution of research is evident, that of policymaking is not as intuitively clear. Yet, policies not only feed and fit into prevalent narratives, they constitute narratives themselves. Rickinson et al. (2018) provide ample comments by (Australian education) policymakers referring to 'narrative' in their work in various forms. Telling a story or constructing a narrative is a primary sense-making tool for humans and often supersedes the persuasive potential of scientific and statistical data (Kahneman, 2011). In addition to the epistemic mode, the narrative-building of meaning is context-oriented (Fischer, 2003). While not following the scientific logic of explaining phenomena, the narrative construction of meaning is not irrational (Fischer, 2003). Policymaking is understood as telling a story that starts with a problem that needs solving, followed by a relevant policy intervention and finally, the outcome (Fischer, 2003) squares to a large extent with our experience as EU policymakers in education and the specific process of preparing the Council Recommendation on learning for the green transition and sustainable development. Here, as is often the case, 'the problem' did not come across as a clear-cut, mono-thematic matter but an amalgamation of similar trends (not necessarily with similar causes) in different countries. The (expected) policy outcome is also a mix: direct action (e.g., through specific projects, funding) as well as more policy (intervention) happening at national, regional, and local levels across the EU.

Looking at the relationship between research and policymaking, there seems to be a gap in terms of researching and presenting the perspective of the policymaker (Shaxson & Boaz, 2021). More often than not, the link between the two is investigated by researchers and with an agenda that is of primary interest to researchers, whereas the

[14] https://ec.europa.eu/transparency/expert-groups-register/screen/expert-groupsconsult?lang=en&groupID=3810

reality and needs of policymakers remain in the dark (Rickinson & al., 2018). This paper tried to shed light on the reality of policymaking in an international context by elucidating the steps leading up to the creation of EU policy initiatives in sustainability education. It outlined the important and extensive role of research in policymaking and, at the same time, reminded us of the fact that it can, by definition, not replace the role of institutional decision-making based on the principle of democracy. In the second part of the paper, I shared my observations with regard to knowledge and research gaps based on the diverse evidence feeding into the proposal and described in the first part of the paper. The paper suggests widening the evidence base with regard to a (i) broader and deeper analysis of policies—educational or otherwise—that affect sustainability education; (ii) local context and its effect on sustainability education policy development and enactment; (iii) improving overall education quality through sustainability education. Each of those themes requires thorough research and appropriate policy attention. To achieve this, researchers and policymakers need to strike a fine balance between rapprochement and critical distance, both of which are necessary to hold up the integrity of their respective spheres of action. Understanding the role of researchers as critical friends seems to fit this purpose best. To be both critical and a friend, it is helpful to get to know each other's field of operation better, including through specific training and exchanges to understand both process and content. At present, 'scientists and policymakers continue to lack regular and well-supported opportunities to meet and exchange ideas' according to the JRC survey among 498 professionals in research and policy (European Commission, 2022g, pp. 16). The WEEC is a pertinent example of a fruitful forum of exchange between research and policymaking. Other formats to foster longer-term exchanges and engagement between researchers and policymakers are dedicated working groups such as the Commission's working group school on learning for sustainability. While it might be hard to quantify the impact of such exchange on either policy or research, judging by our own example, it provides fresh and/or alternative food for thought and avenues for action.

References

Aikens, K., McKenzie, M., & Vaughter, P. (2016). Environmental and sustainability education policy research: a systematic review of methodological and thematic trends. *Environmental Education Research,* 22 (3), 333-359. https://doi.org/10.1080/13504622.2015.1135418

Bianchi, G., Pisiotis, U., & Cabrera Giraldez, M. (2022). *GreenComp. The European sustainability competence framework.* Luxembourg: Publications Office of the European Union. https://data.europa.eu/doi/10.2760/13286

Boswell, C. & Smith, K. (2017). Rethinking policy 'impact': four models of research-policy relations. *Palgrave Commun* 3 (1), 1-10. https://doi.org/10.1057/s41599-017-0042-z

Council of the European Union (2022). *COUNCIL RECOMMENDATION of 16 June 2022 on learning for the green transition and sustainable development* (2022/C 243/01). https://eur-lex.europa.eu/legal-content/EN/ALL/?uri=CELEX:32022H0627(01)

Dixon, D. (2022). *Leadership for sustainability. Saving the planet one school at the time.* Independent Thinking Press.

Dixon, D., Otto, S.; Pensini, P. (2017). Nature-based environmental education of children: Environmental knowledge and connectedness to nature, together, are related to ecological behaviour. *Global Environmental Change* 47, 88-94. http://dx.doi.org/10.1016/j.gloenvcha.2017.09.009

European Commission (2019). COMMUNICATION FROM THE COMMISSION TO THE EUROPEAN PARLIAMENT, THE COUNCIL, THE EUROPEAN ECONOMIC AND SOCIAL COMMITTEE AND THE COMMITTEE OF THE REGIONS. *The European Green Deal.* COM(2019) 640 final. https://eur-lex.europa.eu/legal-content/EN/TXT/?uri=COM%3A2019%3A640%3AFIN

European Commission (2021). COMMUNICATION FROM THE COMMISSION TO THE EUROPEAN PARLIAMENT, THE COUNCIL, THE EUROPEAN ECONOMIC AND SOCIAL COMMITTEE AND THE COMMITTEE OF THE REGIONS. *Better regulation: Joining forces to make better laws.* COM(2021) 219 final. https://eur-lex.europa.eu/legal-content/EN/TXT/?uri=COM:2021:219:FIN

European Commission, Directorate-General for Education, Youth, Sport and Culture, (2022a). *Data collection and analysis of Erasmus+ projects : focus on education for environmental sustainability : final report,* Publications Office of the European Union. https://data.europa.eu/doi/10.2766/29038

European Commission, Directorate-General for Education, Youth, Sport and Culture, Tasiopoulou, E., Billon, N., & Finlayson, A. (2022b). *Education for environmental sustainability: policies and approaches in European Union Member States : final report,* (E. Tasiopoulou, editor, N. Billon, editor, A. Finlayson, editor, H. Siarova, editor, K. Pribuišis, editor, A. Gras-Velazquez, editor, I. Mulvik, editor, M. Bajorinaitė, editor, E. Sabaliauskas, editor, V. Fronza, editor, J. Vežikauskaitė, editor, A. Disterheft, editor) Publications Office of the European Union. https://data.europa.eu/doi/10.2766/391

European Commission, Directorate-General for Education, Youth, Sport and Culture, Looney, J., O'Shea, M., Staring, F.et al. (2022c), *Key competences for all – Policy design and implementation in European school education : final report,* Publications Office of the European Union, 2022. https://data.europa.eu/doi/10.2766/1951

European Commission, Directorate-General for Education, Youth, Sport and Culture, Finlayson, A., Moso-Diez, M., & Orlovic Lovren, V. (2022d). *Impact of COVID-19 on ESD in the context of twin transition,* Publications Office of the European Union. https://data.europa.eu/doi/10.2766/05850

European Commission, Directorate-General for Education, Youth, Sport and Culture, (2022e). *Learning for the green transition and sustainable development: staff working document accompanying the proposal for a Council recommendation on learning for environmental sustainability,* Publications Office of the European Union. https://data.europa.eu/doi/10.2766/02392

European Commission (2022f). *Proposal for a COUNCIL RECOMMENDATION on learning for environmental sustainability,* COM(2022) 11 final. https://eur-lex.europa.eu/legal-content/EN/TXT/?uri=CELEX-%3A52022DC0011

European Commission (2022g). *Staff Working Document - Supporting and connecting policymaking in the Member States with scientific research.* https://knowledge4policy.ec.europa.eu/file/staff-working-document-supporting-connecting-policymaking-member-states-scientific-research_en

Fischer, F. (2003). Public Policy as Narrative: Stories, Frames, and Metanarratives. In Fischer, F. *Reframing Public Policy: Discursive Politics and Deliberative Practices* (pp. 161-180). Oxford: online edn, Oxford Academic, 1 Nov. 2003. https://doi.org/10.1093/019924264X.003.0008

Ginsburg, M. B., & Gorostiaga, J. M. (2001). Relationships between Theorists/Researchers and Policy Makers/Practitioners: Rethinking the Two-Cultures Thesis and the Possibility of Dialogue. *Comparative Education Review, 45*(2), 173–196. https://doi.org/10.1086/447660

Glavic, P. (2020). Identifying Key Issues of Education for Sustainable Development. *Sustainability 2020,* 12, 6500; https://doi:10.3390/su12166500

Kahneman, D. (2011). *Thinking, Fast and Slow.* London: Penguin.

Lingard, B. (2021). The Changing and Complex Entanglements of Research and Policy Making in Education: Issues for Environmental and Sustainability Education. *Environmental Education Research*, 27 (4). 498–512. https://doi.org/10.1080/13504622.2020.1752625

Lysgaard, J.A., Reid, A., & van Poeck, K. (2016). The roots and routes of environmental and sustainability education policy research – an introduction to a virtual special issue. *Environmental Education Research,* 22 (3), 319 - 332. https://doi.org/10.1080/13504622.2015.1108392

Mulà, I., Tilbury, T. (2023). *Teacher education for the green transition and sustainable development – Analytical report*, Publications Office of the European Union. https://data.europa.eu/doi/10.2766/144189

Orland, M. (2009). Separate Orbits: The Distinctive Worlds of Educational Research and Policymaking. In Sykes, G., Schneider, B., & Plank, D. (Eds.), *Handbook of Education Policy Research* (pp. 113–28). New York: Routledge. https://doi.org/10.4324/9780203880968

Rickinson, M., Walsh, L., de Bruin, K. and Hall, M. (2018). Understanding evidence use within education policy: a policy narrative perspective, *Evidence & Policy,* 15 (2), 235–252. https://doi.org/10.1332/17442641 8X15172393826277

Rickinson, M., & McKenzie, M. (2021a). The research-policy relationship in environmental and sustainability education. *Environmental Education Research,* 27(4), 465-479. https://www.doi.org/10.1080/13504622.2021 .1895973

Rickinson, M., & McKenzie, M. (2021b). Understanding the Research-Policy Relationship in ESE: Key Insights from Critical Policy and Evidence Use Literatures. *Environmental Education Research*, 27 (4), 480–497. https://doi.org/10.1080/13504622.2020.1804531

Shaxson, L. & Boaz, A. (2021). Understanding policymakers' perspectives on evidence use as a mechanism for improving research-policy relationships. *Environmental Education Research,* 27 (4), 518-524. https://doi.org/ 10.1080/13504622.2020.1764505

Stratford, R., & Wals, A. E. (2020). In search of healthy policy ecologies for education in relation to sustainability: Beyond evidence-based policy and post-truth politics. *Policy Futures in Education, 18*(8), 976–994. https://doi. org/10.1177/1478210320906656

United Nations Educational, Scientific and Cultural Organization, UNESCO (2019). *Framework for the implementation of Education for Sustainable Development (ESD) beyond 2019*, 40 C/23. https://unesdoc.unesco.org/ ark:/48223/pf0000370215

Chapter 3

Climate change education in the national curriculum framework

why is it so difficult?

Agnieszka Kozłowska[1]

Abstract

Climate change (CC) is one of today's most serious challenges, and climate change education (CCE) is an important tool in combating the climate change crisis. However, despite widespread agreement on the role of education in overcoming the climate crisis, the *Getting Every School Climate Ready* UNESCO report shows that a number of countries have not yet integrated CCE into their national curricula. The question is, why? And what are the barriers to implementing CCE in the national policy and curriculum?

This study aims to analyze the reasons for CCE's absence in the national curricula by looking at the case of Poland, where the national curriculum framework does not include CCE or includes it to a marginal extent. Eisner's null curriculum concept is used as the pedagogical theory. The following five factors that may influence the weak position of CCE in the national core curriculum are discussed: 1) the general attitude of the country towards CC challenges; 2) the role of CCE in climate change policy; 3) the political orientation of the governing party; 4) the level of democracy in the development of the core curriculum; and 5) gender ratio among ministerial posts in government.

Content analysis of policy documents and UNFCCC country reports, supplemented by data from the Climate Change Performance Index and government website, to assess the significance of factors 1

1 Adam Mickiewicz University in Poznan, Poland

and 2. A historical perspective was used to discuss factor 3. The level of public participation in the development of the core curriculum was analyzed to assess the significance of factor 4, and the gender ratio was calculated from government website data.

The findings showed Poland's low performance in CC international commitments and prognosis and an underestimation of formal CCE as a measure for CC mitigation and adaptation. The analysis also revealed a low level of social participation in the legislative process of core curriculum development but no relation between the political orientation of the parties and the CCE position in the national curriculum framework. Women's under-representation in the government structure was another possible factor that might influence the CCE's marginal presence in formal education in Poland.

Keywords: climate change education, education policy, climate change policy, national curriculum, null curriculum

Introduction

Climate change is one of the most serious challenges that humans have ever faced (Giddens, 2011). For the last three decades, the IPCC (Intergovernmental Panel on Climate Change) has been publishing more and more alarming reports on the global warming state of play and its possible consequences, proving that climate change (CC) mitigation and adaptation are a matter of emergency (*Reports — IPCC*, n.d.). In the special report on the impact of global warming of 1,5 C (IPCC, 2022), the Summary for Policymakers indicates education as one of the measures that may "(…) accelerate the wide-scale behavior changes consistent with adapting to and limiting global warming to 1.5°C." (IPCC, 2018, p. 22). This statement enhances the globally spread belief that education is an important tool to mitigate global warming, expressed, among others, in education-focused provisions of international agreements: Art. 6 of the UN Framework Convention on Climate Change (*Convention documents | UNFCCC* n.d.); art. 12 of the Paris Agreement (*The Paris Agreement | UNFCCC*, n.d.), as well as SDG 13 of UN Agenda 2030 (*THE 17 GOALS | Sustainable Development*, n.d.). The importance of education for combating global warming was also echoed by Glasgow's COP26, the first Education and Environment Education Ministers' Summit, on which Audrey Azoulay, UNESCO Director-General said: "There is no solution without education. Every learner needs to understand climate change and be empowered to be a part of the solution, and every teacher is given the knowledge to teach about that. States must mobilize for this" (*Only Half of the National Curricula in the World Have a Reference to Climate Change, UNESCO Warns | UNESCO*, n.d.).

However, although education on climate has been a legal obligation of 197 UN climate change convention's parties for more than 30 years, a surprisingly large number of countries did little or nothing to build knowledge and awareness in the field of CC. In the 2021 UNESCO revision of national curricula from 100 UN countries, partly based on the previous report (*Learn for Our Planet: A Global Review of How Environmental Issues Are Integrated in Education - UNESCO Digital Library*, n.d.), nearly 47% of national curricula had no reference to climate change and the rest usually included the topic in a marginal manner (*Getting Every School Climate-Ready: How Countries Are Integrating Climate Change Issues in Education - UNESCO Digital Library*, n.d.). Moreover, McKenzie's (2021) in-depth study, based on 377 submissions to the UNFCCC secretariat, tracking countries' progress in CC convention commitments, showed that country reports usually lack data suitable for monitoring climate change education (CCE). Only 22% of countries provided some data on Action for Climate Empowerment, and only 2% included data on the SDG indicator 13.3.1, which addressed CC in formal education and national curricula frameworks.

These findings are not a straightforward indication that there is no CCE at schools. There are countries that have implemented CCE and treated it as a priority. In many countries, teachers have autonomy that lets them decide about teaching on CC regardless of the absence or insufficient coverage of this topic in the national curriculum framework. However, the statistics show that the needs and the expectations towards CCE are higher. Too many countries treat CCE marginally, although climate change awareness among the public is necessary to support CC mitigation and adaptation policies and measures (IPCC, 2018).

Contrary to the meager presence of CCE in national curricula, there is public support that advocates for climate change education, including adolescent students, who even riot on the streets protesting because of the slow progress in climate politics and a lack of CCE in schools (*Climate Strike*, n.d.; *Youth Demands for Quality Climate Change Education - UNESCO Digital Library*, n.d.; *Students Want Compulsory Climate Education*, n.d.). Also, parents and teachers expect CCE at schools (*Why Isn't My Kid Learning about Climate Change in Their High School Classroom?*, n.d.; If Climate Change Education Matters, Why Don't All Teachers Teach It?', 2023). In the IPSOS report on teachers' attitudes to climate change, 86% of them agreed that CC should be taught at schools (*Teachers Agree That Climate Change Is Real and Should Be Taught in Schools*, 2019). Moreover, the study on British teachers' opinions on CCE showed preferences for an early and cross-curricular approach to CCE (Howard-Jones et al., 2021).

In light of the above-listed facts related to the discrepancy between the strong need and weak presence of CCE in national curricula, the question of why arises. If there is an international consensus on the importance of CC mitigation and adaptation for CCE, confirmed by binding provisions of international agreements and public pressure to incorporate CCE into formal education, why has there still been so little CCE content in the national curricula documents?

The question also concerns Poland, which, for years, has avoided incorporating CCE into the national core curriculum. CCE is absent at the primary school level of the core curriculum for students aged 7–15 (Kozłowska, 2021a, 2021b) and is only marginally present at the secondary level, mainly in advanced geography in grammar school (*Raport Edukacja klimatyczna w Polsce*", 2021).

In this paper, I use Poland's case to discuss possible explanations for the lack of CCE content in the national core curriculum.

Theoretical background

The theory used in the paper is based on the null curriculum concept by Elliot Eisner (1985), a leading researcher in curriculum reform in the United States. According to Eisner, there are three curricula that all schools teach: the explicit curriculum, the implicit curriculum, and the null curriculum. The *explicit curriculum* is offered to students as an official document with goals, purposes, content, methods of teaching, and assessment. It describes what should or is expected to be taught and learned at schools. The *implicit curriculum* is what is taught alongside the explicit one, a set of rules and expectations communicated "between the lines", in a "hidden" manner, and which "define school as a cultural system that itself teaches important lessons" (Eisner 1985, p. 107). The *null curriculum* embraces options students are not offered, but that is important for the student's future. These are fields or subjects not planned and not taught at schools, neglected by teachers/policymakers regardless of their strong importance for students' future choices, assessments, and decisions based on them. The lack of this content has severe consequences. As Eisner (1985) puts it: "What schools do not teach may be as important as what they do teach". The null curriculum refers to omissions "not a result of ignorance but a product of choice" (p. 97).

Elliot Eisner believes that the national curriculum is seldom a result of professional expertise and well-thought-out pedagogical theory. "Subjects which are taught at school are not a result of careful analysis (…) but rather because they have traditionally been taught (1985, p. 103) (…) The subjects that are now taught are also protected by the interests of teachers who view themselves as specialists in particular fields" (1985, p. 105).

This mix of tradition, politicization of education, and unwillingness to leave the comfort zone that comes from being in a well-known environment leads to a situation in which consecutive national curricula introduced by consecutive governments do not change a lot. However, the world around the school is subjected to a permanent and deep transformation. They remain as they have always been, with the only change being the growing gap between what is taught and socially and economically relevant.

In this study, climate change education is recognized as a part of the *null curriculum*. In answering the question "why?", Eisner's belief is taken into account that a null curriculum is not a coincidence, but the result of particular interests. According to this belief, a delay or resistance in the incorporation of CC content into the national curriculum is a conscious, albeit concealed decision of educational decision-makers.

The core curriculum—a national curriculum framework in Poland

In Poland, the core curriculum, a key document in national education policy, is a long (over 360 p) and detailed description of education goals, content, ways of assessment, methods of teaching, and the education process organization. After a general section containing the general aims of education and the tasks of the school, each school subject at each educational level is described in detail in terms of aims, content, teaching methods, and assessment methods. The core curriculum is compulsory for all schools, and pupils take a nationally uniform external examination at least once in the educational cycle to check their mastery of the content. Teachers can enrich the core curriculum with their own objectives and content, but due to the specificity of this document, most pupils learn according to a set standard.

The development and approval of the core curriculum are regulated by the minister responsible for education. Formally, it is a legislative act—a regulation of the *Education System Act*, which does not have to go through the Parliament, and the Minister decides on the extent of its public consultation. The status of CCE in the core curriculum must then be connected with the government's intentions and characteristics.

Climate change in the core curriculum

The issue of climate change is almost absent from the Polish core curriculum. In my 2021 research, I showed that neither this concept nor its synonyms appear in the primary school core curriculum (Kozlowska 2021 a,b). Since Poland ratified the Climate Convention in 1994 (*United Nations Treaty Collection*, n.d.), the core curriculum has been extensively revised 3 times, and minor curricular changes have also been made several times - but by the time this article was published, climate change had not become an important topic in Polish curriculum documents. At the primary level, it is absent, and at the secondary level, it is placed in extended geography but treated marginally, not negligently.

Research hypotheses

I decided to explore possible reasons for the CCE's absence in the Polish core curriculum about 1) the general attitude of the country towards CC challenges; 2) the role of CCE in climate change policy; 3) the political orientation of the governing party; 4) the level of democracy in the development of the core curriculum; 5) gender ratio among ministerial posts in government.

The possible reasons for the weak position of CCE in the curriculum are as follows:

- (H1) The country's general attitude towards CC challenges is negative, which results in avoidance of CC topics in education policy and core curriculum.
- (H2) The country's CC policy does not recognize education as a measure of CC mitigation or does not transfer it to education policy.
- (H3) The far-right, populist political orientation of the government is associated with CC skepticism and results in the government's conscious avoidance of incorporating CCE into the core curriculum.
- (H4) The limited level of democracy in the procedure of establishing and developing the national core curriculum places decision-making overwhelmingly within the government administration, which has a conservative attitude to climate change.
- (H5) The low proportion of women in the government results in a less ecological orientation of the policy, as women are more eco-conscious than men.

Methods

Document analysis was used as the method to answer the questions. National communication submissions for the UNFCCC secretariat and Climate Change Performance Index were used to assess the country's attitude towards CC challenges; CC policy documents, as well as the above-mentioned country's FCCC reports, were used to assess the perception of CCE in the CC policy; the historical approach was used to assess if political orientation might be a reason of weak CCE position in the national curriculum. The level of democracy was reconstructed based on media information about the degree of stakeholder participation in the process of developing the core curriculum, as well as official government responses to the parliamentary questions on CCE incorporation into the national core curriculum. Gender balance was assessed by the proportion of women in cabinet positions in the government.

According to H2, documents were initially searched by keywords (education, core curriculum, ministry of education) to identify texts for further quantitative and qualitative analysis. The table below illustrates the list of documents, their sources, and the codes, categories, or questions used in the analysis.

Table 1. List of documents analyzed and codes/questions used in the analysis.

Hypothesis	**Documents**	**Sources**	**Codes/categories of information**
H1: The country's attitude towards CC policy	CC Performance Index *Report on the technical review of the fourth biannual report*	UNFCCC secretariat webpage https://unfccc.int/documents	The level of coverage GHG reduction commitments Position on the index
H2: The role of climate change education in climate change policy documents	The National Environmental Policy 2030; National Energy and Climate Plan for the years 2020–2030) *Poland. Biennial Reports (BR). BR 3. National Communication (NC). NC 7. | UNFCCC.* (n.d.). *Poland. National Communication (NC). NC 8. Biennial Reports (BR). BR 5. | UNFCCC.* (n.d.).	Official webpage of the Polish government and Ministry of Environment and Climate mos.gov.pl ww.gov.pl UNFCCC secretariat webpage https://unfccc.int/documents	Participation of the ministry responsible for education in the creation of the documents The role of formal education in CC mitigation Keywords: education, (climate change) education, core curriculum, ministry of education
H3: political orientation and CC denialism	Press releases, media coverage for citizens' initiatives, answers to parliamentary questions	Internet	CC skepticism/denialism
H4: level of democracy of the procedure of the core curriculum construction	Parliamentary questions on CCE in the core curriculum Media coverage of Climate Round Table 2021	Interpelacje, zapytania, pytania i oświadczenia poselskie - Sejm Rzeczypospolitej Polskiej	Attitude towards CCE and CCE in the national core curriculum
H5: Gender balance in the government	Governmental webpages	Rada Ministrów - Gov.pl - Portal Gov.pl (www.gov.pl)	The proportion of female to male government ministers

Results and discussion

Ad. 1 General country's orientation towards CC-associated challenges

Poland is the sixth-largest country, with 38 million people, in the European Union (EU), and it has been a member since 2004. For decades, the country's energy policy (EPP2040, n.d.) has been predominantly based on its sources of coal and gas, resulting in prospective CCS (Carbon Capture and Storage), i.e., "clean coal technology" becoming more economically justified than renewable energy. However, the share of renewable energy in the energy balance was growing and achieved 21.6% in 2020. Since 1994, Poland has been a party to the UNFCCC and is obliged to reduce GHG emissions. The EU policy in this respect is even more demanding, indicating the obligation of 55% GHG reduction towards 1990. As such, there is a strong external pressure to change to a greener course and climate-friendly policies.

To assess Poland's attitude towards CC challenges, performance in the execution of international commitments was taken into account. I chose the UNFCCC commitments on GHG reduction and the position in the CC Performance Index based on the insightful analysis of countries' achievements in CC mitigation from a long-term perspective.

According to Poland's fourth biennial report to UNFCCC, Poland is to gradually reduce GHG emissions in reference to 1990 across almost all sectors. Due to the gradual systemic change of 1989 (the introduction of democracy and capitalism) and the subsequent EU membership since 2004, the progressive modernization of the country has forced an increasingly stringent environmental policy. Between 1990 and 2018, the GHG emission decreases were the biggest in the waste sector (–40.9%), agriculture (–31.7%), energy (–30.6%), manufacturing industries, and construction (–25.9%). However, in the transport sector, the trend was just the opposite. Since 1990, transport emissions increased by 214.4%, which has ruined the effect from other sectors (*Report on the Technical Review of the Fourth Biennial Report of Poland | UNFCCC*, n.d.). Under the Convention, by 2020, Poland should reduce emissions by 20%, but it has achieved only 13.1% so far. Also, the targets from the EU seem difficult to achieve. The European Green Deal target is to make the EU climate-neutral by 2050 and reduce emissions by 55% in reference to the 1990 level (*Delivering the European Green Deal*, n.d.).

Policies on climate and energy, however, introduced many new measures to meet the targets, which, in experts' opinion, might be insufficient to achieve the goals. It is visible in the Climate Change Performance Index, which monitors climate change mitigation efforts of 57 (2021) and 59

(2023) countries plus the EU—covering 92% of the global GHG emissions. The index is built on four pillars: GHG emission (40%), energy use (20%), renewable energy (20%), and CC policy (20%). Each pillar includes several elements, e.g., evaluation of GHG emissions includes the current level of GHG emissions per capita; past trends in GHG emissions per capita; the current level of GHG emissions per capita compared to well-below-2°C compatible pathway; GHG emission reduction 2030 compared to well-below-2°C-compatible pathway (Burck et al., 2019). According to the last CCPI reports from 2020 and 2023, Poland is placed at the bottom of the ranking with the category of a "very low performer," and its result is worsening (Burck et al., 2019, 2022), which is shown in Tables 3 and 4. According to these reports, Poland, in its policy, has no neutrality goal and, at the same time, is one of the nine countries responsible for 90% of global coal production.

Table 3. Top CCPI performers in 2020 and 2023.

	2020			2023	
Rank	**Country**	**Score**	**Rank**	**Country**	**Score**
4	Sweden	75.77	4	Denmark	79.61
5	Denmark	71.14	5	Sweden	73.28
6	Morocco	70.63	6	Chile	69.54
7	United Kingdom	69.80	7	Morocco	67.44
8	Lithuania	66.22	8	India	67.35
9	India	66.02	9	Estonia	65.14
10	Finland	63.25	10	Norway	64.47
11	Chile	62.88	11	United Kingdom	63.07
12	Norway	61.14	12	Philippines	62.75
13	Luxemburg	60.91	13	Netherlands	62.24
14	Maltha	60.76	14	Portugal	61.55
15	Latvia	60.75	15	Finland	61.24
16	Switzerland	60.61	16	Germany	61.11
17	Ukraine	60.60	17	Luxemburg	60.76
			18	Malta	60.42

Source: (Burck et al., 2019, 2022)

Table 4. Bottom CCPI performers in 2020 and 2023.

2020			2023		
Rank	**Country**	**Score**	**Rank**	**Country**	**Score**
48	Turkey	40.76	50	Japan	40.85
49	Bulgaria	40.12	51	China	38.80
50	Poland	39.98	52	United States	38.53
51	Japan	39.03	53	Hungary	38.51
52	Russian Federation	37.85	54	Poland	37.94
53	Malaysia	34.21	55	Australia	36.26
54	Kazakhstan	33.39	56	Malaysia	33.51
55	Canada	31.01	57	Chinese Taipei	28.35
56	Australia	30.75	58	Canada	26.47
57	Islamic Republic of Iran	28.41	59	Russian Federation	25.28
58	South Korea	26.75	60	South Korea	24.91
59	Chinese Taipei	23.33	61	Kazakhstan	24.61
60	Saudi Arabia	22.03	62	Saudi Arabia	22.41
61	United States	18.60	63	Islamic Republic of Iran	18.77

Source: (Burck et al., 2019, 2022)

The data shows that Polish CC policy and politics are not sufficient to meet the international commitments on GHG reduction, both under UNFCCC and under the European Green Deal. It shows a rather negative than positive attitude towards CC challenges: CC policy is present, but it seems it is not the state's priority.

Is low performance on climate targets the reason for the lack of climate education in the core curriculum? Such a relationship seems natural. On the other hand, there are countries, like the UK, which have treated CCE marginally in the national curriculum (Greer et al., 2021; Kozłowska, 2021a), but their politics on CC places them between top performers in the CCPI (see Tab 3.). Some countries, like Canada or Australia (Fahey, 2018; J. Fahey et al., 2014), are bottom performers in the CCPI (Table 4) but have implemented CCE education at schools. However, it seems plausible that CCE is more effective regarding the country's climate policy and politics.

Ad. 2. Perception of CCE in climate change policy documents

Education is marginally treated in all the documents analyzed. In the National Environmental Policy 2030, the section on education is a two-page passage entitled "Environmental education, including behavior patterns for sustainable consumption" (pp. 77-78) and "Intervention's direction: Environmental education, including behavior patterns for sustainable consumption". In these chapters, building and enhancing the environmental awareness of Polish citizens is shown as a strategic activity of the state, which should be executed through formal, informal, and non-formal education, with participation, cooperation, and dialog with all important actors: scientists, NGOs, and citizens. On page 111, in the part of an action plan for the strategy, the obligation is to write *The National Strategy for Environmental Education,* but the deadline for this task is not until 2025. The only minister responsible for this task is the minister responsible for the environment. This suggests that the task is not treated as urgent or envisaged as part of the formal education system.

In the document, education is mainly a tool to support sectoral activities in the green transformation of waste management, transport, and energy. All but one educational tasks are attributed to the ministry responsible for the environment. The ministry responsible for education is mentioned only once in the whole document assigned to the general task of formal environmental education and awareness raising. Education, although mentioned 54 times in the document, does not occur in the context of climate change, with the one exception that educational activities for sustainable consumption were known to be desirable precisely in the context of climate change. The term "climate change", mentioned as many as 220 times in the document, does not appear in the context of education, although much space is devoted to descriptions of the expected impacts of climate change and the needed mitigation and adaptation measures. In the documents analyzed, climate change and education 'do not see each other': education is not considered a useful tool in the actions designed to address the climate crisis. Previous research has shown that the education system does not see climate change as an important topic of education.

Table 3. Results of content analysis of the environmental and CC policy documents.

Keywords	Number of Occurrences/Context		
	Total	National Environmental Policy 2030	National Energy and Climate Plan for years 2021–2030
Education/ climate change education/ environmental education	63	54 Developing environmental competencies (knowledge, skills, attitudes) for sustainable consumption; the need for formal, informal, and non-formal education in that regard; general statements on the need for rising environmental and sustainability awareness The need of environmental education needed in greening sectors of transport and energy The obligation of building an environmental education strategy until 2025	9 Education of the general public on CC causes, effects, and CC mitigation and adaptation; training of medical personnel on new illnesses connected to CC; training for inhabitants of regions at risk of CC effects; training for farmers on the CC-connected risks; building system of states programs supporting citizens affected by CC effects; the need of formal and informal education on CC; environmental education for state programs on electromobility, green transport, and energy
Minister/ Ministry of education	2	1—leading and enhancing formal education in the fields covered by the Environmental Policy, including cooperation with universities and NGOs	1—mentioned once at the beginning of the document as an executive minister but not assigned to any specific task in the following parts of the text

In the second document—National Energy and Climate Plan for the years 2021–2030— environmental education is seen as a tool to support climate change adaptation activities, but in very limited contexts. It refers to the training of medical personnel on new infectious diseases related to climate change, training farmers in the context of climate change impacts and how to obtain compensation for natural disasters, and inhabitants of areas more exposed to the negative impacts of climate change. It also talks about the need for more climate education for the general public and potential beneficiaries of government programs related to the green transition. None of these tasks, however, are assigned to the Minister of Education, although he is mentioned at the beginning of the document as one of the implementers of the policy. The Ministry of Education is virtually absent from this document and excluded from the long list of the 13 institutions that contributed to the above documents.

The next two documents, country submissions to the UNFCCC secretariat (national communications number 3 and 5 from 2028 and 2022, respectively), are a government elaboration on progress in implementing the provisions of the Climate Convention, including Article 6 committing the state to climate change education. I was looking for information on the involvement of the Ministry of Education in the preparation of the report and a list of climate education activities in the country.

The older 2018 document was prepared with the cooperation of nine ministries and six government-associated institutions. The Ministry of Education or any of its subordinate institutions, e.g., for teacher training, was not a cooperating party. The newer report from 2022 was prepared as a cooperation of four ministries and five institutions. Again, there were no Ministry of Education representatives working on the document. In both documents, a chapter on CCE (an obligation under Article 6 of the Convention) focused rather on general environmental education, not specifically CCE, as well as informal education provided by the Ministry of Environment and Climate, National Forests or the National Found of Environmental Protection and Water Management, not the Minister of Education.

The weak position of CCE in CC policy documents may come from a lack of recognition of CCE as an important measure of CC mitigation, but also from the general perception of education as an unimportant sector of politics. There is no national strategy document on education similar to the National Environmental Policy 2030 or Energy and Climate National Plan for the years 2021–2030, although such documents exist at a lower administrative level (local authorities) for cities, communes, and regions (*Program Rozwoju Edukacji w Warszawie w Latach 2013–2020 | Społeczna Strategia Warszawy*, n.d.; 'Strategia rozwoju oświaty 2015–2020', 2023). At the state level, there is only a short, two-page document on the official webpage of the Ministry of Education, entitled "Basic directions of the national education policy in the year 2022/23" (*Podstawowe kierunki realizacji polityki oświatowej państwa w roku szkolnym 2022/2023—Ministerstwo Edukacji i Nauki—Portal Gov. pl*, n.d.). On the 10-point list, there is no reference to CCE, sustainability, or environmental issues.

Summing up, albeit marginally, environmental and CC policy documents include education as a measure in CC policies. However, there is no explicit reference to formal education, with its vital role in improving environmental awareness, instilling values, and shifting attitudes systematically for the whole population of children and youth. Obligations of the Climate change convention are implemented through action-based, informal education, and the Ministry of Education is not included in either the plans for writing the Environmental Education Strategy or the reports to the UNFCCC in relation to Article 6.

Ad.3. Political orientation of the governing party

To check if political orientation could be the reason for the weak position of CCE in the curriculum, a historical perspective was used to show environmental and CC education in the context of the political orientations of consecutive governments.

The current version of the core curriculum, prepared by the *Law and Justice* government, is the third one since Poland's most comprehensive education reform in 1999. After the collapse of the communist system in 1989, a set of ambitious reform programs was conducted by the government of Jerzy Buzek. A three-level territorial organization of the state (government, territorial authority, local authority) was carried through; together with administrative reform, the educational system was reorganized and decentralized. Schools started to be led by local authorities, and the core curriculum was introduced, with modernized and depoliticized content, new school subjects, and more freedom for teachers to create their school-level curricula (Wiśniewski & Zahorska, 2020).

Except for traditional school subjects such as math, physics, biology, history, or geography, inter-subject educational paths were introduced for environmental education, health education, new media, life in society, and at the higher level of schools—philosophical, regional, and European education. The curriculum content of environmental education was supplemented with the National Strategy for Environmental Education, "Education for Sustainable Development". The document was initiated by cooperating ministers responsible for education, environmental protection, and natural resources and forestry, and created in wide cooperation of government administration, experts, and NGOs, who in 1995 organized huge environmental conferences for all actors in the process of writing the national strategy. In 1998, the Strategy was accepted by committees of environmental protection of lower and higher chambers of the Polish Parliament and once again improved and actualized before being published in 2001. The Strategy was planned for ten years and was connected with another strategic national document—the Environmental Policy of the State (*Narodowa Strategia Edukacji Ekologicznej | Centrum Edukacji Ekologicznej Ełk*, n.d.). The last document stated that:

"Environmental education is one of the most effective methods of investment into the change of environmental awareness of the society. These changes should be treated as fundament to the success and social acceptance of transformation towards sustainable development" (p. 25).

The National Strategy for Environmental Education, together with the environmental inter-subject, cross-curricular path in the national curriculum, was an execution of the provisions of the Constitution, the environmental protection bill, and the national education system bill. The last one obliged public authorities to promote environmental knowledge among children and youth, assure environmental safety, and support citizens in their efforts to protect the environment.

Ten years later, in 2009, Donald Tusk's government decided to revise the core curriculum. The content was reduced. Teachers received more autonomy to create their curriculum and collect suitable teaching materials. However, inter-subject paths disappeared; environmental education was again spread and incorporated into main school subjects, like primary science, biology, and geography. The National Strategy for Environmental Education, prepared for 2001–2010, was not continued nor replaced by the new document.

Eight years after the core curriculum revision, in 2017, the far-right populist government of Mateusz Morawiecki decided to conduct another deep structural and curriculum reform of the education system. The changes diminished school democracy and teachers' autonomy. The new core curriculum content was changed into a more traditional and detailed one.

In the new curriculum, CCE was absent at the primary level (7–17 years old) and had a marginal position in the curriculum for the secondary school level, being limited to non-obligatory, advanced geography and only one chapter in the curriculum for vocational schools.

This absence was a part of the wider problem of rather superficial and fragmentized environmental education in the Polish national curriculum, focusing on the benefits of water and energy saving at home, as well as materials recycling, but ignoring the environmental crisis, climate change, ecological transformation, and carbon footprint as an important reference point in the environmental impact assessment (Kozłowska, 2021b).

Table 4. History of changes to the national curriculum framework against the background of changes in government and its political orientations

Time	Prime minister	Term of the Parliament	Governing political party	New versions/ fundamental reconstruction of the national curriculum framework
1997–2001	Jerzy Buzek	III	AWS—Solidarity Electoral Action (Center-right)	1999, the first *Core curriculum*—a regulation of the new Education System Act Minister responsible for the reform: prof. Miroslaw Handke
2001–2004	Leszek Miller	IV	SLD—Democratic Left Alliance (Center-left)	
2005–2006	Kazimierz Marcinkiewicz	V	PiS—Law and Justice (Right-wing populism)	
2006–2007	Jaroslaw Kaczynski	V	PiS—Law and Justice (Right-wing populism)	
2007–2011	Donald Tusk	VI	PO-PSL (Center-right/ Christian democracy)	2009 The first revision of the *Core curriculum*; minister responsible for the reform: Katarzyna Hall
2011–2014	Donald Tusk	VII	PO-PSL (Center-right/ Christian democracy)	
2014–2017	Beata Szydło	VIII	PiS—Law and Justice (Authoritarian right-wing populism)	2017—The new *Core curriculum.* Minister responsible for the reform: Anna Zalewska
2017–2019	Mateusz Morawiecki	IX	PiS—Law and Justice (authoritarian right-wing populism)	
2019–	Mateusz Morawiecki	X	PiS—Law and Justice (authoritarian right-wing populism)	

Source: (*Polska. Historia, Encyklopedia PWN*, n.d.), own elaboration.

Might the government's political orientation be responsible for the CCE position in the curriculum? On the one side, right-wing populist parties and politicians are closer to climate change denialism (McCright & Dunlap, 2011; Kulin et al., 2021; Laferrière & Stoett, 1999; Stott, 2021). Skepticism of political elites about scientific climate predictions, combined with the imperative of consumption-driven development, is a mixture that leads to a lack of effective climate policy. This attitude, characteristic of conservative, populist governments, can be camouflaged by ostensibly pro-climate rhetoric, the construction of successive climate change strategies, and government participation in international agreements, as it was described by Young & Coutinho (2013). However, at least for part of the public, such a policy is fully acceptable, as political orientation influences perceptions of climate change not only among political elites but also among citizens (Czarnek et al., 2021; Hamilton, 2011; McCright & Dunlap, 2011).

The far-right political orientation of the current government might explain why there is no quality environmental or CC education in the new core curriculum. However, except for the cabinet of Jerzy Buzek, previous governments, although far from right-wing populism, also paid little attention to climate education.

Summing up, it needs to be noted that despite many examples, including a Polish one, of conservative governments not supporting effective climate policy or education policy covering climate topics, there are no arguments that left or center-oriented governments always treat CCE as a priority.

AD.4 The level of democracy in the curriculum establishment and development

Democratic lawmaking is a strong supporter of CC mitigation policy and education. In Singapore, the consultative procedure of creating educational policy that involves experts' opinions saved CCE in a curriculum, as they introduced this content into documents they consulted even though government officials omitted the topic in the original proposal (Chang & Pascua, 2017).

In Poland, formally, higher-level legislative projects such as bills should be submitted to public consultations, discussed and voted on in the Parliament. However, the core curriculum is a lower-lever legislative document that constitutes the sole responsibility of the minister of education, who ultimately decides on the depth and scope of social participation in the legislative decision-making process.

The extent to which the core curriculum is consulted significantly affects its quality and subsequent implementation. By being kept informed and given the opportunity to make comments, teachers, parents, and other interested parties accept the changes and actively participate in their implementation. The core curricula published in 1999 and 2009 were widely consulted. However, the level of participation referring to the last national curriculum reform in 2017 was very low. The change of the core curriculum was not based on evidence and was carried out amid protests by teachers, supported by parents and students (Wiśniewski & Zahorska, 2020). Social consultancy of the regulation was shortened to a minimum; the whole procedure was hidden from the eyes of the public and evoked controversies. The curriculum Authors' names were not public, regardless of many official appeals to reveal them. After two years of battle, the administrative court forced the ministry responsible for education to reveal the names—most of them were unknown in the education science community (*Reforma edukacji. Sąd nakazał MEN ujawnienie nazwisk autorów podstawy programowej*, 2017). The controversies and protests were due to the arbitrary and politicized content choice, including some excessive additions or serious omissions, the latter being the lack of climate change topics.

Even after the implementation of the new core curriculum, protests and calls for climate education continued. Official calls on the Minister of Education came from the members of Parliament (*Interpelacja nr 14192 do ministra edukacji narodowej w sprawie treści dotyczących zmian klimatycznych oraz ochrony środowiska w podstawie programowej nauczania w szkołach podstawowych i ponadpodstawowych*, 2020; *Interpelacja Nr 20542— Tekst—Sejm Rzeczypospolitej Polskiej*, n.d.) and university professors (*Naukowcy apelują o wprowadzenie edukacji klimatycznej. Niezwłocznie*, n.d.; *„Poloniści wobec zmian klimatu" | Raport naukowców z UŚ | Uniwersytet Śląski w Katowicach*, 2021), as well as youth organizations. In 2021, The Climate Strike, Extinction Rebellion, WWF, UN Global Compact Network Poland and other climate change-oriented NGOs organized a Climate Round Table on the necessity of CCE in the core curriculum. The organizers invited the ministers responsible for the environment, climate, and education, to participate in the event. Although none of the invited ministers appeared, under traditional and new (social) media pressure, they eventually signed a final agreement on introduction of CCE into the national curriculum by 2023. Apart from this announcement, however, not much has happened to make the declaration reality. Still climate change education, including webinars for teachers, is offered by the Ministry of Environment and Climate, but not the Ministry of Education. There were no legislative changes in the core curriculum, nor massive preparation on teachers' education in this field.

AD.5. Gender (dis)proportion in the government.

Gender differences in environmental attitudes are well-recognized and verified. In a Swiss study on cantons referenda, women supported environmental projects more often than men (Funk & Gathmann, 2015). In a Swedish study on a large sample of 2400 students aged 12 to 19, a gender gap in sustainability awareness was found. Girls were more environmentally aware than boys, and the discrepancy grew with age and was higher in schools with environmental profiles (Olsson & Gericke, 2017). In the literature review in a 10-year period, 1988–1999, using the New Ecological Paradigm (NEP) Scale to study the environmental awareness of adults, women were more aware than men (Zelezny et al., 2000), to mention just a few of the studies on gender aspects in ecology. Socialization of gender roles is a probable mechanism of these differences. Girls are educated to care for and boys to compete (Holemans et al., 2022). Another explanation is the psychological perception of genders. Ecology, linked more with care than competition, is perceived as unmanly (Brough et al., 2016). There are also gender differences in a hierarchy of values: altruism, essential in an environmental attitude, is more important to women than men (Dietz et al., 2002). Another factor that influences gender differences in environmental attitudes is risk perception. Women are more sensitive to risk and perceive the same situation as more threatening than men do. In the context of the climate change crisis, women perceive environmental threats as more serious (Bord & O'Connor, 1997). For these reasons, the gender balance in the government may play a role in the general government's attitude to climate change policy, including education.

At the date of the closing of this article, the Council of Ministers comprised 28 members; ministerial functions were held by 24 men (83%) and 4 women (17%).

Conclusion

Climate education is a prerequisite for overcoming climate change. However, while most of the world agrees with this fact and has officially accepted the obligation of climate education as a party to the UNFCCC, only a few countries have written climate education into their national programmes. They are implementing it as part of formal education.

Poland is majority ignoring climate change education, even though the state is a party to the Climate Convention, a member of the European Union. It has developed national ecological and climate policies, and its citizens are actively demanding environmental education within the formal, universal education system. Unfortunately, bottom-up pressure was insufficient to change the situation: members of the national parliament, scientists, and a coalition of NGOs have unsuccessfully called on the Ministry of Education to include climate change education in the national curriculum. It seems that there is no CCE in the Polish curriculum because there is no political will to do so and because there is a legal possibility to determine the content of the core curriculum arbitrarily—by a single decision of the Minister of Education and a narrow group of selected social partners.

In this respect, the analysis confirms that climate change education is part of a *null curriculum*—consciously omitted part of the national education standard.

The question of why it is so difficult to put climate education on the national agenda requires further research based on a large comparative sample of many countries. However, the case study reviewed here provides a glimmer of an answer. Climate education should be seen as part of and a consequence of the overall national policy toward the climate change challenge. Stopping climate change requires more than a low-carbon economy: everyone's life should become low-carbon, which requires a change in environmental awareness, behavior, and the whole culture (Gola, 2022). Behavioral change, if carried out on a sufficiently large scale, could make a significant contribution to reducing GHG emissions (Cordero et al., 2020). This task is unattainable without education. If governments really wanted to promote behavioral change among citizens, they would make climate education a priority. In the case presented, this is not happening.

The lack of climate education in the Polish core curriculum is consistent with the poor results of climate policy to date and the low-ranking position in the Climate Change Performance Index. The persistent resistance of those in power to the introduction of climate education, despite the appeals of civil society and experts, shows that the lack of climate education in the formal education system is a conscious decision by the authorities, who "actively ignore facts", the mechanism described by Young & Coutinho (2013) and McIntyre (2018). In the analyzed case, the situation is not changed by the educational activities of the Minister for the Environment and Climate, as he has no competence in the area of formal education. Climate education carried out by ministries other than education has an action-oriented, ad hoc character and, by definition, covers only narrow groups of interested parties. In the analyzed case, the Ministry of Education was not invited to contribute to national environmental policy or climate policy documents. This fact was true of all four analyzed documents, so it can hardly be considered a coincidence. It suggests that education and environmental policies do not see each other; the national core curriculum does not see climate change and environmental crisis, and environmental policy documents do not recognize education as an effective tool for their targets.

The limited democratic nature of the approval process for the core curriculum reinforces the effect of a lack of political will motivated by climate change skepticism. This effect overlaps with gender imbalance in government, with a very low ratio of women, statistically more progressive and pro-climate than men.

The reasons for the CCE's absence in the national core curriculum discussed in the study do not exhaust the list of possibilities. For example, interdisciplinarity and the high level of complexity of a climate change phenomenon make it difficult to fit into traditional subjects, curricula and teacher competencies. Climate change can also be difficult for teachers who are experts within narrow disciplines and may not feel competent enough to teach about climate change. As a new topic, climate change may also be insufficiently covered by didactic support materials. Lessons on this topic would require more effort from teachers, etc.

There seem to be many possible barriers to climate change education, which should be and are studied. What was shown in this paper is that climate change education is very sensitive to political influences. The issue is politicized, which should be taken into account in climate education research, as this provides additional context and a source of potential determinants of the presence and quality of climate education in a country.

References

Bord, R. J., & O'Connor, R. E. (1997). The Gender Gap in Environmental Attitudes: The Case of Perceived Vulnerability to Risk. *Social Science Quarterly*, *78*(4), 830–840.

Brough, A. R., Wilkie, J. E. B., Ma, J., Isaac, M. S., & Gal, D. (2016). Is Eco-Friendly Unmanly? The Green-Feminine Stereotype and Its Effect on Sustainable Consumption. *Journal of Consumer Research*, *43*(4), 567–582. https://doi.org/10.1093/jcr/ucw044

Burck, J., Uhlich, T., Bals, C., Niklas Höhne, & Nascimento, L. (2019). *2020 CCPI Climate Change Performance Index*. GermanWatch New Climate Institute. https://chttps://ccpi.org/download/the-climate-change-performance-index-2020/cpi.org/wp-content/uploads/CCPI-2023-Results-3.pdf

Burck, J., Uhlich, T., Bals, C., Niklas Höhne, & Nascimento, L. (2022). *2023 CCPI Climate Change Performance Index*. GermanWatch New Climate Institute. https://ccpi.org/wp-content/uploads/CCPI-2023-Results-3.pdf

Chang, C.-H., & Pascua, L. (2017). The Curriculum of Climate Change Education: A Case for Singapore. *Journal of Environmental Education*, *48*(3), 172–181.

Climate Strike. (n.d.). Retrieved 2 March 2022, from https://www.climatestrike.net/

Convention documents | UNFCCC. (n.d.). Retrieved 15 September 2022, from https://unfccc.int/process-and-meetings/the-convention/history-of-the-convention/convention-documents

Czarnek, G., Kossowska, M., & Szwed, P. (2021). Right-wing ideology reduces the effects of education on climate change beliefs in more developed countries. *Nature Climate Change*, *11*(1), Article 1. https://doi.org/10.1038/s41558-020-00930-6

Delivering the European Green Deal. (n.d.). [Text]. European Commission - European Commission. Retrieved 3 June 2022, from https://ec.europa.eu/info/strategy/priorities-2019-2024/european-green-deal/delivering-european-green-deal_en

Dietz, T., Kalof, L., & Stern, P. C. (2002). Gender, Values, and Environmentalism. *Social Science Quarterly*, *83*(1), 353–364. https://doi.org/10.1111/1540-6237.0088

Eisner, E. W. (1985). *The educational imagination: On the design and evaluation of school programs* (2nd ed). Macmillan ; Collier Macmillan.

Energy Policy of Poland until 2040 (EPP2040)—Ministry of Climate and Environment—Gov. pl website. (n.d.). Ministry of Climate and Environment. Retrieved 31 May 2023, from https://www.gov.pl/web/climate/energy-policy-of-poland-until-2040-epp2040.

Funk, P., & Gathmann, C. (2015). Gender gaps in policy making: Evidence from direct democracy in Switzerland. *Economic Policy*, *30*(81), 141–181. https://doi.org/10.1093/epolic/eiu003

Fahey, S. J. (2018). Curriculum change and climate change: Inside outside pressures in higher education. In *Curriculum and Environmental Education*. Routledge.

Fahey, S.J., R. Labadie, J., & Meyers, N. (2014). Turning the Titanic: Inertia and the drivers of climate change education. *Journal of Applied Research in Higher Education*, *6*(1), 44–62. https://doi.org/10.1108/JARHE-01-2013-0003

Giddens, A. (2011). *The politics of climate change* (2nd ed., rev. updated). Polity Press.

Getting every school climate-ready: How countries are integrating climate change issues in education—UNESCO Digital Library. (n.d.). Retrieved 9 March 2022, from https://unesdoc.unesco.org/ark:/48223/pf0000379591

Gola, B. (2022). Education for Ecological Values. How to Build the Bond between Children and Nature? *Edukacja Elementarna w Teorii i Praktyce*, *17*, 95–107. https://doi.org/10.35765/eetp.2022.1767.07

Greer, K., King, H., & Glackin, M. (2021). *The 'web of conditions' governing England's climate change education policy landscape. Journal of Education Policy*, 1–24. https://doi.org/10.1080/02680939.2021.1967454

Hamilton, L. C. (2011). Education, politics and opinions about climate change evidence for interaction effects. *Climatic Change*, *104*(2), 231–242. https://doi.org/10.1007/s10584-010-9957-8

Holemans, D., Osman, P., & Fransen, M.-M. (2022). *Dare to Care: Ecofeminism as a source of inspiration*. SKRIBIS. https://gef.eu/publication/dare-to-care/

Howard-Jones, P., Sands, D., Dillon, J., & Fenton-Jones, F. (2021). The views of teachers in England on an action-oriented climate change curriculum. *Environmental Education Research*, *27*(11), 1660–1680. https://doi.org/10.1080/13504622.2021.1937576

If Climate Change Education Matters, Why Don't All Teachers Teach It? (2023, March 3). *Education Week*. https://www.edweek.org/teaching-learning/if-climate-change-education-matters-why-dont-all-teachers-teach-it/2023/03

Interpelacja nr 14192 do ministra edukacji narodowej w sprawie treści dotyczących zmian klimatycznych oraz ochrony środowiska w podstawie programowej nauczania w szkołach podstawowych i ponadpodstawowych. (2020, November 9). https://www.malgorzatatracz.pl/interpelacja-nr-14192-do-ministra-edukacji-narodowej-w-sprawie-tresci-dotyczacych-zmian-klimatycznych-oraz-ochrony-srodowiska-w-podstawie-programowej-nauczania-w-szkolach-podstawowych-i/

Interpelacja nr 20542—Tekst—Sejm Rzeczypospolitej Polskiej. (n.d.). Retrieved 19 February 2023, from https://www.sejm.gov.pl/Sejm9.nsf/InterpelacjaTresc.xsp?key=BYSK88

IPCC. (2022). Global Warming of 1.5°C: IPCC Special Report on Impacts of Global Warming of 1.5°C above Pre-industrial Levels in Context of Strengthening Response to Climate Change, Sustainable Development, and Efforts to Eradicate Poverty (1st ed.). Cambridge University Press. https://doi.org/10.1017/9781009157940

IPCC, 2018: Summary for Policymakers. In: Global Warming of 1.5°C. An IPCC Special Report on the impacts of global warming of 1.5°C above pre-industrial levels and related global greenhouse gas emission pathways, in the context of strengthening the global response to the threat of climate change, sustainable development, and efforts to eradicate poverty [Masson-Delmotte, V., P. Zhai, H.-O. Pörtner, D. Roberts, J. Skea, P.R. Shukla, A. Pirani, W. Moufouma-Okia, C. Péan, R. Pidcock, S. Connors, J.B.R. Matthews, Y. Chen, X. Zhou, M.I. Gomis, E. Lonnoy, T. Maycock, M. Tignor, and T. Waterfield (eds.)]. Cambridge University Press, Cambridge, UK and New York, NY, USA, pp. 3-24. https://doi.org/10.1017/9781009157940.001.

Kozłowska, A. (2021a). Climate Change Education in Polish and UK National Curriculum Frameworks. *Lubelski Rocznik Pedagogiczny*, *40*(3), 103–118. https://doi.org/10.17951/lrp.2021.40.3.103-118

Kozłowska, A. (2021b). Edukacja ekologiczna w polskiej podstawie programowej. *Rocznik Pedagogiczny*, *44*, 123–150. https://doi.org/10.2478/rp-2021-0009

Kulin, J., Johansson Sevä, I., & Dunlap, R. E. (2021). Nationalist ideology, rightwing populism, and public views about climate change in Europe. *Environmental Politics*, *30*(7), 1111–1134. https://doi.org/10.1080/09644016.2021.1898879

Laferrière, E., & Stoett, P. J. (1999). *International relations theory and ecological thought: Towards a synthesis*. Routledge.

Learn for our planet: A global review of how environmental issues are integrated into education—UNESCO Digital Library. (n.d.). Retrieved 15 March 2022, from https://unesdoc.unesco.org/ark:/48223/pf0000377362

McKenzie, M. (2021). Climate change education and communication in global review: Tracking progress through national submissions to the UNFCCC Secretariat. *Environmental Education Research*, *27*, 1–20. https://doi.org/10.1080/13504622.2021.1903838

McCright, A. M., & Dunlap, R. E. (2011). The Politicization of Climate Change and Polarization in the American Public's Views of Global Warming, 2001–2010. *The Sociological Quarterly*, *52*(2), 155–194. https://doi.org/10.1111/j.1533-8525.2011.01198.x

Narodowa Strategia Edukacji Ekologicznej | Centrum Edukacji Ekologicznej Ełk. (n.d.). Retrieved 16 February 2023, from http://cee.elk.pl/dokumenty/narodowa-strategia-edukacji-ekologicznej

Naukowcy apelują o wprowadzenie edukacji klimatycznej. Niezwłocznie. (n.d.). Rzeczpospolita. Retrieved 19 February 2023, from https://klimat.rp.pl/ekotrendy/art19289071-naukowcy-apeluja-o-wprowadzenie-edukacji-klimatycznej-niezwlocznie

Olsson, D., & Gericke, N. (2017). The effect of gender on students' sustainability consciousness: A nationwide Swedish study. *The Journal of Environmental Education*, *48*(5), 357–370. https://doi.org/10.1080/00958964.2017.1310083

Only half of the national curricula in the world have a reference to climate change, UNESCO warns | UNESCO. (n.d.). Retrieved 21 May 2023, from https://www.unesco.org/en/articles/only-half-national-curricula-world-have-reference-climate-change-unesco-warns

Podstawowe kierunki realizacji polityki oświatowej państwa w roku szkolnym 2022/2023—Ministerstwo Edukacji i Nauki—Portal Gov.pl. (n.d.). Ministerstwo Edukacji i Nauki. Retrieved 31 May 2023, from https://www.gov.pl/web/edukacja-i-nauka/podstawowe-kierunki-realizacji-polityki-oswiatowej-panstwa-w-roku-szkolnym-20222023

Poland. Biennial Reports (BR). BR 3. National Communication (NC). NC 7. | UNFCCC. (n.d.). Retrieved 27 May 2023, from https://unfccc.int/documents/198284

Poland. National Communication (NC). NC 8. Biennial Reports (BR). BR 5. | UNFCCC. (n.d.). Retrieved 27 May

2023, from https://unfccc.int/documents/624748

Poloniści wobec zmian klimatu" | Raport naukowców z UŚ | Uniwersytet Śląski w Katowicach. (2021, September 28). https://us.edu.pl/polonisci-wobec-zmian-klimatu-raport-naukowcow-z-us/

Polska. Historia, Encyklopedia PWN: Źródło wiarygodnej i rzetelnej wiedzy. (n.d.). Retrieved 31 May 2023, from https://encyklopedia.pwn.pl/haslo/Polska-Historia;4575043.html

Program rozwoju edukacji w Warszawie w latach 2013-2020 | Społeczna Strategia Warszawy. (n.d.). Retrieved 19 February 2023, from http://spolecznastrategia.um.warszawa.pl/content/program-rozwoju-edukacji-w-warszawie-w-latach-2013-2020

Reforma edukacji. Sąd nakazał MEN ujawnienie nazwisk autorów podstawy programowej. (2017, September 22). wyborcza.pl. https://wyborcza.pl/7,75398,22413611,reforma-edukacji-sad-nakazal-men-ujawnienie-nazwisk-autorow.html

Raport „Edukacja klimatyczna w Polsce". (2021, July 14). https://ungc.org.pl/raport-edukacja-klimatyczna-w-polsce/

Reports—IPCC. (n.d.). Retrieved 21 March 2023, from https://www.ipcc.ch/reports/

Report on the technical review of the fourth biennial report of Poland | UNFCCC. (n.d.). Retrieved 28 May 2023, from https://unfccc.int/documents/278828

Stott, P. (2021). *Hot air: The inside story of the battle against climate change denial.* Atlantic Books.

Strategia rozwoju oświaty 2015-2020. (2023, January 2). *Oświata w Czerwonaku.* https://oswiataczerwonak.pl/strategia-oswiatowa/strategia-rozwoju-oswiaty-2015-2020/

Students want compulsory climate education. (n.d.). Retrieved 21 July 2023, from https://theecologist.org/2021/sep/28/students-want-compulsory-climate-education

Teachers Agree That Climate Change is Real and Should Be Taught in Schools. (2019, April 22). Ipsos. https://www.ipsos.com/en-us/news-polls/teachers-agree-climate-change-real-and-should-be-taught-schools-04-22-2019

THE 17 GOALS | Sustainable Development. (n.d.). Retrieved 28 May 2023, from https://sdgs.un.org/goals

The Paris Agreement | UNFCCC. (n.d.). Retrieved 1 May 2022, from https://unfccc.int/process-and-meetings/the-paris-agreement/the-paris-agreement

UNFCCC. (1992). United Nations Convention on Climate Change. Retrieved 15 May 2023, from https://unfccc.int/resource/docs/convkp/conveng.pdf

United Nations Treaty Collection. (n.d.). Retrieved 2 October 2023, from https://treaties.un.org/Pages/ViewDetailsIII.aspx?src=IND&mtdsg_no=XXVII-7&chapter=27&Temp=mtdsg3&clang=_en

Why isn't my kid learning about climate change in their high school classroom? (n.d.). MIT Climate Portal. Retrieved 21 May 2023, from https://climate.mit.edu/ask-mit/why-isnt-my-kid-learning-about-climate-change-their-high-school-classroom

Wiśniewski, J., & Zahorska, M. (2020). Reforming Education in Poland. In F. M. Reimers (Ed.), *Audacious Education Purposes* (pp. 181–208). Springer International Publishing. https://doi.org/10.1007/978-3-030-41882-3_7

Young, N., & Coutinho, A. (2013). Government, Anti-Reflexivity, and the Construction of Public Ignorance about Climate Change: Australia and Canada Compared. *Global Environmental Politics, 13*(2), 89–108. https://doi.org/10.1162/GLEP_a_00168

Youth demands for quality climate change education—UNESCO Digital Library. (n.d.). Retrieved 28 May 2023, from https://unesdoc.unesco.org/ark:/48223/pf0000383615

Zelezny, L. C., Chua, P.-P., & Aldrich, C. (2000). New Ways of Thinking about Environmentalism: Elaborating on Gender Differences in Environmentalism. *Journal of Social Issues, 56*(3), 443–457. https://doi.org/10.1111/0022-4537.00177

Chapter 4

Art and ethnology in the service of environmental education for a sensitive anchoring in the service of resilience to the Anthropocene

Edith Planche

As an ethnologist, I founded the association SeA Science et Art in 2000, which implements environmental education through culture, i.e., art and ethnology, in partnership with local authorities, schoolchildren or children in leisure time and adults. My book, *Eduquer à l'environnement par l'approche sensible*, reflects this double trajectory with a first anthropological part and a second pedagogical part, including implementation workshops linking theory and practice. The first anthropological part hypothesizes that the ruptures between Science–Art and Nature–Culture are of the same epistemological strain at the origin of the disconnection between the human and his environment. The second part proposes a pedagogy of the sensitive, where the in-itself of the emotional and intimate subject, the at-home of the daily, and the local subject can approach the general and the conceptual, which also corresponds to a reversal of the paradigm of objectivity underlying the Nature–Culture and Science–Art ruptures. The proposed methodology invites building bridges between human and nature, between each individual and his environment in the broad sense, and between the experiences of each one and the learnings. The subtitle, "Art, ethnology and ecology" evokes bridges woven between different disciplines. Between art and environmental education, ethnology and environmental education (ethnography of the territory, ethnopedagogy to address other paradigms, anthropology to shift the focus on the nature–culture relationship), and art and ethnology.

We will show how art and ethnology, by contributing to restoring sensitive links to help environmental education, can develop the culture of virtuous circles rather than cultivating a virtue based solely on guilt. Environmental education through sensitivity re-energises the subject on many levels, promoting autonomous thinking to incorporate commitment that makes sense. Let's become "sensiblocene"*, a profound driver of behavioural change. (*term coined by myself corresponding to the lack of the sensible that makes the world lame…)

Keywords: environmental education, sensitive, nature-culture, art, ethnology, enchantment

Art and ethnology in the service of environmental education for a sensitive anchoring in the service of resilience to the Anthropocene

Cascading bridges[1]

As an ethnologist, I founded the SeA Science et Art[2] association in 2000, which implements environmental education through culture[3], i.e., art (culture) and ethnology (cultures) in partnership with local authorities, schools, children in leisure situations and adults, thus creating bridges between research and action, between anthropology and educational approaches. Indeed, the genesis of this ESD methodological approach was forged from an anthropological observation. At the dawn of the 21st century, despite the eruption of uncertainty and complexity (Morin, 2000, 2003), the call for a non-Cartesian epistemology (Bachelard, 1934), and the milestones of plural postmodernity (Sauvé, 1999), positivist[4] thought is still trailing the comet of our European societies structured by the separation of the subject from the object, the sensible from the intelligible, and leading Man in techno-scientific societies to extract himself from himself and his subjectivities, to separate himself from nature and the objects of the world.

In terms of restoring links, the structure of SeA, Science and Art, aims to reintegrate the sensitive into learning by encouraging crossings between the analytical spirit and the poetic breath. The ESD educational actions consider art as a tool of cognition as well as a tool of mediation and mediatisation according to the motto "Create to love and learn", "Create to love and preserve", "Create to 'demonstrate' by showing" (Planche, 2007). The actions carried out encourage bridges between languages (intelligible and sensitive, scientific and artistic) and between generations, cultures and Nature and Culture to raise awareness of the challenges of ESD.

The bridges between concepts and application are illustrated by the concretization of these principles on the local terrain of the Rhône in an educational programme[5] conducted with 20,000 primary school pupils in eight departments along the river for more than 15 years, in partnership[6] with the local authorities, ministries and rectorates concerned. This project invites each class along the river to explore its local river with the educational tool *Le Rhône, la mallette des merveilles du fleuve*[7], and then to create a collective artistic portrait for each class, giving rise to regional exhibitions, bridging the gap between the school and the region. The kit proposes to study the natural heritage from the perspective of cultural heritage through an *art and science* dynamic[8]. In addition, this programme builds intergenerational bridges by integrating the participation of families[9] and fellow citizens via sensitive memory-sharing walks and mobilising all the school's ages around a common project.

The action made it possible to explore the river by reenchanting it, by giving it volume through art and ethnography, for an appropriation of the problems of water and its uses at the heart of the territory by structuring a feeling of belonging to the issues of the Rhône in proximity in perspective of the river as a whole and finally, of the question of water in the world, thus building bridges between local and global experience on different scales: sensitive experience (creative intimacy, daily experience, and local experience) and learning (water, energy, fauna and flora, regional planning, landscape, transport, waste, air quality, climate, and human activities) referring to transversality between the different school disciplines in France (history, geography, biology, technology, visual arts, French, etc.).

These experiences are conceptualised in my book *Eduquer à l'environnement par l'approche sensible* (Planche, 2018), which reflects this double anthropological and educational trajectory, with a first anthropological part and a second pedagogical part also including implementation workshops bridging theory and practice. The first section postulates that the separation between 'science'[10] and 'art' is of the same epistemological strain as the separation between 'nature' and 'culture', and that these divorces are part of a prevailing paradigm of rupture at the root of the disconnection of humans from their environment. These sections examine the cultural underpinnings of these separations in the light of other ontologies in European geographical space and time. The second section proposes workshops to build a pedagogy of sensitivity, most often experimented concretely within the SeA Science and Art projects and via its kit. This methodology of environmental education through sensitivity invites us to build bridges between humans and nature, between each individual and their environment in the broadest sense of the term—the natural environment as well as the living environment—and between the experiences of each individual and learning. The subtitle Art, ethnology and ecology indicates that the book builds bridges between various disciplines. Between art and environmental education, between ethnology and environmental education (ethnography of the territory, ethnopedagogy (Pardo, 2002) to address other paradigms, anthropology to shift the gaze on the nature-culture relationship) but also between ethnology and art for a re-enchanted[11] methodological approach.

Indeed, the transdisciplinarity of the actions is expressed in this generic back-and-forth between *art and sciences*[12] that runs through each environmental education tool, whether it is themed in nature (ecosystems, fauna, flora, etc.) or culture (economy, history and ethnography of the territory). In addition, the frequentation of the works of art masters,[13,] as well as the artistic reflex, intervened as supports to the conception of the tools, in the form of transdisciplinary reasoning,[14] making the bridge between sensitivity and reflection. My personal practice of painting[15] has familiarised me with inductive and analogical thinking, but also with the attitude of "controlled drifting" which presides over the artistic exercise to induce dilations favourable to creativity before the phase of shaping (drifting familiar to the ethnographic exercise in the field to favour impregnation, before the analysis phase). As for anthropology, it intervenes at three levels :

- At the meta-level, through a societal analysis of the cascading ruptures that structure rationalist epistemology to highlight forms of educational epistemological colonialism.[16]
- My professional practice in ethnology in France, which led to the publication of individual and collective works (Planche, 1989 and 1999, cited by Guy Vincent 1992, 1996), plunged me into the heart of the cultural

identity issues characteristic of the end of the nineteenth and beginning of the twentieth centuries, in the face of a loss of anchorage to the local, natural and/or cultural heritage. This scientific experience made it possible to design the ethnographic workshops in the book (Planche, 2018)[17]

- Finally, the anthropological analysis of the dominant epistemologies in different societies associated with the pedagogical question allowed me to propose ethnopedagogical[18] workshops expressed exoterically through art and poetry.

We will show how the development of environmentally sensitive hooks can sketch answers to the questions problematised by Bob Jickling on the foundations of the moral impulse and the need to act in the face of environmental issues: "*Should ethics be based on duties? Can it be joyful? Can effective ethics be based on abstract principles, or must they be born of lived experience?* "[19] Environmental education through culture, i.e., art and ethnology, makes it possible to develop an approach of sensitive anchoring to the territory and to knowledge that creates bridges with oneself and with the environment in the broad sense (natural environment as a living environment) to promote an incorporated responsibility based on the virtuous circles of meaning.

In addition to the syndrome of the lack of nature and links, there is a dimension of loss of meaning that the re-enchantment of reality, the playfulness, and the symbolic volume of our responses to the world can alleviate and make it possible to question the paradigms of our relationship to the environment and reality in this context of the Anthropocene. We will examine how the bridges between the artist, the animist, the ethologist and the ethnologist can restore volume, fullness, life and vitality to our relationship with the world.

The characteristic of this proposal is to provide answers through art and ethnology to the question of *what to do, concretely, in a profane way, in a modern society such as ours*. Art and ethnology can relay the *wild thought*[20] of which we can be lame, for a sensiblocene becoming in the service of the resilience to the anthropocene.

From anthropology to pedagogy

The objective of developing a concept to encourage the decompartmentalization of "Art" and "Science"[21] (in the sense given to them by European culture) was outlined with the anthropological observation of the separation between intelligible and sensible, between substance and form, and around the question of the cultural basis of these separations[22]. I founded the association SeA, Science and Art, in line with these anthropological reflections, to reconcile these poles in an educational form that would combine "critical spirit" and "poetic breathing" to rehabilitate culture and the sensitive as heuristic and fertile.

This objective was in line with the concerns of Edgar Morin, godfather[23] of SeA, to recognise the heuristic place of the Humanities with the desire to build bridges not only between the so-called hard sciences and the Humanities and Social Sciences but also between Art and Science, with in particular the importance of literature in thinking about the notion of subject[24]. (Morin, 1984, 2000).

The hypothesis that the separations between Nature and Culture, Science and Art, are the two interfaces of the same cleavage at the source of the distancing of Man from his environment invites us to examine the paths of convergence between rationalism and rationalisation. The cleavage between "Science" and "Art", implemented with structural rationalism, which cuts us off from the subject, from the symbol and from a more holistic, even volumetric, way of thinking, is of the same strain as the rupture between "Nature" and "Culture".

implemented with the rationalisation of spaces, which cuts us off from the ground and turns nature into an object to be used coldly. It is these separations that motivate the attempts at connection advocated by complex thinking:

> "*The supporters of classical science saw in this manipulative power offered to men by science only the proof of man's progress, moving from the stage of his adaptation to the environment to that of the adaptation of his environment. This quest to tame the world has led, we noted with E. Morin, to the forgetting of the complex unity of man and nature to the profit, if there is profit, of a deplorable dichotomy between the universe of objects and that of subjects, between the objective and the subjective, between man and nature.*" (Banywesize, 2007, p.31)

While the objectivist paradigm corresponding to the Science–Art break asks us to separate ourselves from the subject and the sensitive in order to better observe reality (Descartes invites us to be wary of the trade of the senses and sensitive misguidance) by positioning ourselves *outside of ourselves*; the Nature–Culture break, which leads us to think of the world as an object—a resource to be exploited, relies on technology and the urbanisation[25] of spaces in order to situate us *outside of the ground.*

On the scale of geographical space, Descola (2008) states that the concept of nature as an object is *'far from being shared by all the peoples of the planet, who are attached to other cosmological principles'*. As for the time scale, our models of rural societies in the 19th and early 20th centuries remained closely linked to the nature of proximity, which conditioned survival (Planche, 2011). At the beginning of the 20th century, the sociologist Max Weber showed how the technicisation of everyday life was part of this march towards the disconnection between man and his environment. Today, the weather or the ripening of cheese are measured by machines. Modern humans are losing their ability to build bridges of meaning with the clues of the ecosystem to predict rain and to use their senses linking body and know-how—smelling and tapping the wheel of cheese to indicate the stage of maturation—as the traditional craftsperson did. Technology and the "all-soil" way of life have distanced humans from a direct link with their senses[26] and with nature.

It is this basis of rationalisation that makes brutal and sometimes irreversible cuts in the ecosystem, that mutilates landscapes[27] and animals with grids and cages[28], that puts minds into the mould of single-mindedness and top-down planning against the fine composition of the local terrain. Thinking and spaces are subjected to univocal norms.

Thus, while the rupture between science–art and science–sensitive is part of the ideal model of the techno-scientific society where it is a question of separating oneself from the sensible to remain objective and observe the world, opposing the emergence of the *Subject*, excessive rationalisation unties and loses the *Singular* by providing a generalist response, detached from the link to the local context, which will be applied for everything and everyone.

Therefore, it is a question of the loss of the subject and the singular in the face of an *Objetisation*[29] and *Objectification* (negation of singularity) of the world. It seems that this negation of the subject leads to a negation of the *Pacha-mama*[30] in the sense of a living land and Earth.

This dominant culture leads us to a paradigm of ruptures at the source of the Anthropocene. The Nature–Culture rupture is a part of this seizure of power of positivist epistemology, which conditions our relationship with reality... and underpins the Science–Art rupture to make us evolve in a sort of 'aplatocene', disenchanted world. We are also cut off from the symbol and a fructifying thought in volume. All the profound dimensions of art that open by saying "more" than what it shows, all the literature capable of grasping the subject and the imperceptible of its sensitive convolutions, do not seem a priori necessary[31] to build this functional techno-scientific world. The separation between Science and Art, which is disenchanting (the notion of enchantment containing the idea of psalmody, of vibration, as per Max Weber), is of the same strain as the rupture between Nature and Culture, which conveys values of coldness, of "flatness" and of distance.

De-rationalising the world or re-establishing bridges?

We hypothesise that resilience to the Anthropocene must involve de-rationalisation, i.e., re-establishing the paths of the logic of nature in the face of the rationalisation of spaces and reinjecting the crossroads of the sensitive and the subject in the apprehensions of reality in the face of structural 'rationalism'.

Today, it is a question of re-establishing connections and bridges:

- At the level of spaces to rediscover the connected logic of nature, for example, as advocated by permaculture.
- At the level of thought patterns, by giving voice to the ramifications and digressions of sensitive links that take side roads, a nod to the vocabulary of Chthulucene (Haraway), which advocates a new narrative with multi-connections and the ability to interweave different realities[32] (as opposed to Cartesianism, which urges us to un-weave, disentangle).

We will advocate the "de-rationalisation" and/or "re-enchantment" of our relationship with reality in order to reconcile Culture and Nature (for example by recreating green and blue grids to re-establish the logic of Nature, which can be translated into SFN[33] replacing the notion of SRN); recreate bridges between Man and Nature, between Art and Science, between the cold sign and the fertile symbol, between the rational and the imaginary, reason and feeling. For example, with the sensitive walks with a tourist or educational[34] aim, we will be interested in what links each person to the territory by proposing ethnographic entries (sharing of memories) or creative breaks (art). In the absence of remembering, we leave room for the expression of impressions, associations of ideas "that make me think of", weaving a link between oneself and the environment to be known in order to better digest the learning, whether it is historical-cultural—a walk in the street—or scientific-natural—a walk near a stream! It is a question of re-establishing links with nature but also with the walls, for a reanimated real world! A reconnection.

The aim is not to castigate the rational and the reasoning but to restore the sensitive and life-giving links (which refer us to something other than the functional) with the "objects" of the world in a dominant context of the artefact which meets its critical point with the Anthropocene. It is all a question of re-establishing bridges, of reconnecting:

- With knowledge (as opposed to an instruction based on abstraction and heteronomy). With the sensitive approach, we wetten learning with personal values (the creative, everyday, local subject) instead of drying it up with abstract notions to be integrated through understanding, which, more often than not, can remain far from the concerns of young people.
- With nature (as opposed to the Nature–Culture divide and cladism as a 'posture of observing the rest of the world through the human gaze') with its attendant dominations. Wilderness pedagogy (Jickling et al., 2018), in its theoretical foundations, raises the need to "*disrupt the persistent ideology of human control in relation to the importance of land-based relationships and the possibility of cultural change through education*".[35]
- With the sensible (as opposed to the domination of the intelligible which separates the subject from its sensible perceptions, of which we are lame and as we are lame from "wild thought").

It is, therefore, a question of repairing these cascading divorces, including the rupture reproduced in education between knowledge and the subject, to promote a more holistic education to "*learn to connect knowledge to life*" (Morin, 2014). In the field of education, transmission based on the cumulative and cerebral model condemns the learner to adapt to formatted thoughts that they will most often experience as a heteronomous input, not corresponding to any internal link, nor to the mobilisation of the subject and the living within him or her. In addition, the playful entries do not spare this principle, sometimes contenting themselves with illustrating a question-answer model in a logical sequence.

Contrary to Cartesianism, which asks us to separate ourselves from sensible experiences and untangle what might be confused or mixed up in order to reach clear and distinct ideas, and to that culture of numbers which leads us to distance ourselves from emotions: "*Out of fear of love, philosophers and scientists have directed us towards a culture of the molecule, a culture of numbers, a 'de-affect' culture*" (Cyrulnick cité par O'Dy, 1993), our whole pedagogy is based on developing sensitive attachments to learning and to the environment. This education aims to rebuild bridges between knowledge and the subject, between the subject and his or her environment in the broadest sense, whether through Art which activates creative intimacy around a theme, through filiation to the territory via ethnography which revives an emotional and living memory to reconnect the learner to his or her local history—whether it is from here or abroad—through the mobilisation of the body and the senses which rehabilitates the complete human being, through the meaning given, i.e., the connection to lived experience and life (the daily and local experiences).

This education through sensitive, based on inductive thinking—by starting from oneself and one's experience—will awaken the authorial[36] subject, by mobilising the *self* of the emotional, intimate, and artistic subject, the *at home* of the daily subject and of the local territorial experience, in order to make sense of the general and the conceptual, which also corresponds to a reversal of the paradigm of objectivity that underlies the Nature–Culture and Science–Art divides.

Culture at the service of ESD

ESD through Culture, i.e., Art and ethnology will promote the restoration of bridges with the environment in the broad sense (with nature, the soil, but also with the symbol,—with the objects of the world through Art when following the paths of the multiplication of colors and shapes—with local cultures and territories of life). The sensitive approach, by developing reliance, makes it possible to give meaning for an incorporated and affective appropriation of the ecological transition so that care becomes a sensitive and above all not moral evidence (morality conveys cultural unthought and rigidities)

Art, lever of engagement: Before studying a technical work of the river territory like the bridge, we will invest personal values by creation, with our desires (*designing a basketball bridge, a candy bridge*[37]) and with our imagination (*draw a snake bridge: the sensation reflects a questioning -> The snake bridge moves, what material is it made of?)* to stimulate the desire to look into engineering. We thus participate in the resuscitation of places and environments by reinhabiting the environment with living things (here affect, personal concerns) in the image of animist societies. Or, we will go through the anecdotal, that is to say from life, to arouse the interest of the children in the revision of the functions of the characters of the territory: *for example, draw up their identity card with their tic of language, their shopping list, their favorite CD, their nickname (we will see the footballer Mbappé appear…) mixed with information to acquire on the bargee, the winemaker or the cheesemaker, etc.*[38] , to establish a personal link with learning and the environment to know, in order to invest it with oneself to want to take care of it.

Art, a lever for re-enchantment, develops a holistic way of thinking that restores the link and the "fullness" of reality in volume. Thus, by encouraging a less binary way of thinking, Art makes it possible to inscribe concepts by other means than intellectual understanding: *we will visualise energy sobriety, which will be revealed with the variations of the harlequin's costume*[39], marking the notions of energy saving by the form and the colour, *or we integrate the notion of food chain by a chimera,*[40] which reflects the concept or *work on the identity of the traditional figures by being inspired by Dali's* Don Quixote.[41] The work serves to think about the future and to project oneself, going further, saying more than what it shows and the conscious intention. The artistic and creative approach extends the reflection through the tentacles and the bridges[42] that it launches with the invisible[43], the unspeakable and the possible.

In a landscape reading, the Fauvist art movement will be used to create shocks on the nature-culture representations through the inversion of colours[44] (informative art through a less binary language) *or to express one's moods linked to the landscape through Fauvist colours*[45] *before making a landscape analysis* (engaging art).

Ethnography, a lever for commitment: Becoming aware of one's place in a local culture is a lever for putting down roots so as not to want to damage it anymore, but it is also a vector of visualisation that will make it possible to become aware of the systemic nature of the issues starting from the local territory in order to build bridges with the global level, in a back-and-forth movement between the local and the planetary-global experience Ethnography is therefore a lever for commitment through attachment to the territory but also a lever for bridging the scales from local to global.

However, ethnography is also a lever for re-enchanting the territory for a less binary way of thinking, by putting back the fullness, the link and the volume: *materializing through a drawing all the universe that is expressed in the traces, the etymologies (names of streets, squares, places, bridges) with the workshop entitled "Sleeping Beauty"*[46], it is to make the landscape alive and inhabited by the memory of the elders who forged the territory, that is to say to re-animate the landscape.

Thus, ethnography can awaken the etymologies of place names, while art materializes them. But bridges will also be built between art and ethnology when *the portrait*[47] *of the river, for example, is born with the results of the ethnological survey. The constraints of drawing the morphology are indicated upstream in relation to the questions and answers*, to leave all the room for the imagination to illustrate them... This participates in the re-enchantment and the setting in volume of the sociological analysis and the invested landscape.

Ethnology-anthropology: the intercultural approach makes it possible to provoke a fertile decentring for reflection, to become aware of the Nature-Culture relationship in different societies. The notion of Nature- Culture, among others, is at the heart of the paradigm reversals put into perspective by anthropology (cf. Descola ideal-type, 2005, p 323). The learner will build bridges between here and elsewhere (other societies) but also between past, present and future. (European rural time of the 19th and early 20th centuries) to compare their way of life with other ways of situating themselves in their environment, to relativise their habits and think about ESD. For example, the notion of responsibility and ethics will be enriched and reflected upon in the light of cultural paradigms. It is not a question here of developing avoidance strategies in a "renegotiation of the constraint", but of becoming aware of the preciousness of the territory but also of the object, of the gesture of the manufacturer by comparison with other models (when in certain civilisations, but also in our rural peasant societies, the waste is not discarded but preciously recycled). A workshop will propose *to meet a professional*[48] *with the ethnography of the craft* to establish the link between the object and its maker. If the maker is not identifiable (industrial manufacture), *the object can always be given back its preciousness by working on memory and associations of ideas (by enhancing the affects around the object*[49]*: personal and/or cultural affects in the image of Malinoski's Canoe trobriandais).* When *we associate this work with a collective workshop on the exchange of objects and memories*[50]*, it allows us to value recycling and social links* while circumventing the idea of the frustration associated with the abandonment of the Kleenex gesture, going in the direction of this joyful sobriety...

Ethnopedagogy (Pardo, 2002): It invites us to experience a different relationship with Nature that reconnects with the living, to experience a more connected relationship with the environment via the tools of our society such as poetry **(Art)** (Planche, 2007), a form of "wild thought" (Planche, 2017) or animistic in the generic sense that will promote paradigm shifts for more responsibility and commitment.

This approach to ESD through Art and ethnology thus makes it possible to reduce the distances between oneself and the outside world (the environment in the broad sense, including the perception of reality and "worldly objects" but also the know). The notion of "responsibility" invoked as a slogan (even an advertising slogan) of the good session for the planet can nevertheless be limited to self-righteousness. Art and ethnology will help us to become aware of the preciousness of the environment and of the objects of the world by giving them back volume, by anchoring responsibility in personal and local incorporation that makes sense, going in the direction of "*'integral responsibility', which includes several characteristics of a reconstructivist modernity*" (Sauvé, 1999) for a resilience sensitive to the Anthropocene.

Art enables one to activate a form of "connected" thinking through forms and colours, which link holistically with the wider environment. Moreover, the creative artistic approach is a vector of meaning and pleasure for motivation when one's environment (landscapes, themes, objects) is full of creativity in which one has put personal values. Art mobilises the living in oneself "*away from philosophies rooted in abstract principles and subject-object dualisms. [...] exploring ... ways of experiencing the world that are restorative, affirmative, particular, empathetic and joyful. [...encouraging multiple forms of knowing and doing that restore our embodied sensibilities'*[51] to motivate engagement with the care of environment. Thus, with the artistic approach, one is inclined to establish an intimate and augmented relationship with "the objects of the world" through the fructification of images, i.e., to re-enchant the environment, to re-give volume to living in order to fight against this feeling of evolving in an aplatocene civilisation, empty of depth (the meaning of myth and symbol) and meaning! The ethnological approach allows us to reconnect with the memory of the territory and therefore makes it readable (via the affective narrative: this trace reminds me of my grandfather who used to tell us that he used to take people from one bank to the other in times of war), and connects it to its roots: linking up with the elders as a symbolic filiation gives it meaning and density, which helps us to not degrade it any further. By revealing this bundle of animated experiences that have forged the territory, ethnography gives thickness to the living environment, giving it consistency in the sense of Bruno Latour's recommendations (2018). Behind my street is a filiation that forms symbolic threads of a relationship to the neighbourhood, to the land. Not everything is functional and flat; my neighbourhood, my city, is teeming with memory, inhabited by living words and affects that make it precious, animated and not empty. The artistic and ethnological approaches have in common the capacity to re-establish links and bridges, but also meaning, life, volume and density.

Unveiling the epistemological bridges between the artist, the animist, the ethologist and the ethnologist for a profane wild thought

After COVID-19, the ESD movements that want to "get out" to fight against the syndrome of lack of nature are accelerating. This syndrome is also a symptom of a lack of naturalness and vitality. We are adding a dimension of loss of meaning that the re-enchantment of the world can compensate for, and that allows us to revisit the paradigms of our relationship with the environment in the broadest sense.

What are we missing? How can we become less lame? Less limping?

When we go out into the open air, the invigorating air and the animal and plant encounters will catch up with us and reconnect us to the ground, the wind[52], and the earth. In addition, ethnography will reconnect us to "its" land (to the terroir, whether we are from here or elsewhere, as an anchor to memory). But we also have a great need for the symbolic dimension of looking at the world, for this re-enchantment, for holistic or non-binary thinking, for this augmented experience with the environment and the objects of the world, for getting out of the unidirectional human and 'objective' gaze imposing rigid categories that prevent us from seeing anything else, in order to connect ourselves to the circle of points of view of the Other of the world (Planche, 2022). This Other is wild thought[53], but it is also the non-human points of view to which it gives a platform for an expanded vision. Art allows us to broaden the vision: is art a way of reinjecting wild thought in a profane way?

Today, it is a question of establishing subtle bridges between more analytical forms of thought where "*we evacuate our sensations to manipulate concepts*" and forms of thought that come under the heading of savage thought in the sense of Lévi-Strauss, which is not the thought of savages[54] but a form of thought "*which calculates not with abstract data, but with the teaching of sensitive experience: smells, textures, colours*" (quoted by Sorman, 1989). For we are deprived of a connected and sensitive way of thinking to decipher reality. With the help of Georges Devereux (2010), who states that every society has a primary matrix and a secondary or repressed matrix, we hypothesise that poetry is a secondary matrix of our society that expresses the repressed 'wild thought' of which we are lame. Devereux allows us to consider models of relationship to nature (for example, Descola, 2005, p. 323) as fruitful ideal types for thinking and organising differences while relativising the inscription of these ontologies in a relationship to a specific society. All these modes of presence in the world run through every society more or less emergently. Nevertheless, on a global scale, Western thought and animist thought in a generic sense are opposed according to the following scheme (Planche, 2011 and 2018)[55]:

- For the techno-scientific society, nature is a resource available to humans and must be adapted to their needs. The model of Western societies presents a pyramidal relationship where humans master Nature seen as a silent object–resource, not subject to point of view and intentions.
- The animist model presents a circular relationship: each living, each mountain and river is a subject with a point of view generated by their perceptive organs, their "ecosystems-*conatus*"[56]. To this extent, everything comes alive. Nature is talkative, and each element of the living becomes a subject:

plants, animals, and places have intentions. They are subjects that exchange and confront subjects.

As D. Abram says, "The earth has fallen silent". (Abram, 1996/2021). Should we silence the human voice and path in order to better listen to the dynamics of the ecosystem, to give voice to the discourse of the people autochton[57] who relay the *Pacha-Mama*, this living earth inhabited by all its guests with their points of view and perceptive "enunciation" postures? Because in this Western vision of the world, by being above everything, by considering the rest of the world as a resource, man deprives himself of this unknown (the blind spot of the paradigm of Nature with its needs, its conatus) and excludes himself from the symphony of the living. Poetry seems to be an essential tool for slipping into this "other of the world" (Planche, 2022). Nature is voiceless because it is not a subject, and we should, therefore, be silent to hear its secrets. In an ESD pedagogical approach, we introduce every project with the *Balade poétique* (Planche, 2008, 2011), where we propose to the children to walk in nature in silence to try to capture the secrets of the elf[58] or the goblin by awakening all their senses. It is, in fact, a question of silencing one-way thinking to open up the vision and leave the blinkers of cladism and human understanding behind. Poetry seems to be the culmination of this necessary paradigm shift: it invites us to stop thinking in categories to connect to something other than ourselves. And this is indeed the direction of the Chthulucene (Haraway, 2016). It is with a posture of receptive listening and not of predator that classifies that the poet apprehends the world: poetry is not only a subjective projection but also an encounter with the world (Haldas, 1977), with that Other that our narcissism is reflected.

> "*Didn't Plato mean the same thing to us by evoking a beauty in itself, that is, contemplated and not appropriated or discussed, a beauty not subject to the narcissistic projections of the human?*" (Planche, 2018, p. 124) "*Eternal Beauty that knows neither birth nor death, that never changes. This Beauty does not present itself as a face or as a bodily form, nor is it a reasoning or a science. This Beauty exists in and of itself, simple and eternal; from it, all beautiful things flow. If life is worth living, it is at this moment when the human being contemplates Beauty in itself.* (Platon quoted by Planche, 2018, p.124)

The poet contemplates reality and, by the same token, leaves it a place free of human discourse. Like the animist who recognises that each element of nature has a specific and different point of view, the poet connects with his environment by leaving it its autonomy, by developing a gaze that is silent on explanations, a gaze free of any labelling, to "enter the world", to get rid of the glasses that export the universe of Homo sapiens onto the rest of reality, a nod to the movements for the rights of Nature in the 21st century, which aim to give it back its intrinsic value, outside the needs and projections of the human gaze! For the artist, it is a question of letting himself be touched, or even guided, by the world around him in a relationship of reciprocity and exchange, as André Marchand shows us: "*I believe that the painter must be pierced by the Universe and not want to pierce it*" (Quoted by Laplantine, 1996). The artist will allow themself to be penetrated by the environment that guides and inspires them, thus operating a paradigm reversal for a Western-style "wild thought". Rimbaud's "*I is another*" (1975. Letters of 1871) speaks to us of this intuition stemming from a relationship–perception of the environment that informs, forms,[59] and deforms. With the poetic attitude in the generic sense of Art, the relationship with reality is paradoxically more direct[60] because it dispenses with the "knowledge" that plans the directed look to remain free and contemplative.

The ethologist, the ethnologist, the animist and the artist: bridges for a new paradigm

Thus, we can see bridges between the animist and the artist and today between the animist and the ethnologist when the latter acts as a relay for wild thought. Ethnology teaches us, for example, via the voice of Jean Malaurie (2015), that the Inuit do not want to record their knowledge as hunters because they are wary of words that 'freeze thoughts' (Planche, 2019). Rules can indeed 'freeze intuition'... To be a good hunter, one must remain attuned to this sharpened perception of nature and its clues—just like the approach of the Facteur Cheval[61], an artist of art brut who enters into this ecosystemic relationship with the clues of nature when he affirms that the wind, the birds, and his path guide him to build his work. "You must not demand too much understanding; that would make the stone fall from his hands", whispers the stumbling block that inspired him to start his Ideal Palace. It is indeed this posture that the sensitive approach favours when the learner is invited to de-rationalise his apprehension of the environment.

Poetry paradoxically suggests that we should be silent about words (at least about classification) to "let ourselves be seized" and enter into friendship with Nature by letting go of a directed will and gaze. As the poet Arthur Rimbaud wrote (1975. Letters 1871): "*One should say one thinks me*". Thus, the posture of pedagogy through sensibility is not to know nature but to recreate impressionistic links with nature and the environment. "It is a question of extracting ourselves from our knowledge built around categories in order to re-hear" (Planche, 2019). It is not easy to discard the reflexes of Cartesian teaching that seeks to reintegrate learning into the apprehension of nature instead of espousing the paradigm reversal to let the relationship with the environment live, that is to say, first to "naturerer" itself. Only then, in the second phase, can one use this experience in Nature as a lever for learning. This posture is familiar to the ethnologist when he meets Inuit societies whose social structures are conditioned and shaped by the presence of an extreme natural environment that forms 'hypersensoriality' and sharpens intuitive and creative thinking: Jean Malaurie developed the concept of anthropogeography (Cf. Planche, 2019) of a *naturante* and structuring nature for society.

Moreover, frequenting the posture of the poet, inspired by nature's movements by entering into the ecosystemic resonances and links, in a posture of receptive listening and not of pre-emption, familiarises us with other forms of relationships to nature that the ethnologist brings to light. The ethnologist shows us traditional societies that seem to be in tune with this ecosystemic homeostasis, when the principle of precaution towards nature is not linked to morality but to the fact that the natural elements, objects and subjects of the world are entangled[62], anticipating here the lessons of Chthulucene, who has the intuition that animism is to be combined with the future. (Cf. Haraway, 2016)

It is indeed the poet's posture that opens the way to decentering. As Rimbaud shows us, the artist throws off his categories disanthropocentres himself in order to enter into contact with the world differently when he confides to us, "I is an Other", to allow himself to be grasped rather than constructed with a directed will, in the image of the Inuit, who allows himself to be informed by his sensitive presence at the moment of Nature's movements. The "horribly dull" poetry castigated by Rimbaud (1975. Letters 1871) refers to the notion of a poetry that is thought instead of "let yourself think", because, Dubuffet reminds us, "Art does not come to sleep in the beds that have been made for it. Its best moments are when it forgets what it is called. (Duvaleix, 2005). Platon (Œuvres complètes, 1970) sheds light on this when he evokes the studious poet who is a failed poet because he has withdrawn from the "delusions of the muses" (Cf. Planche, 2017, p. 72), that is to say, from that inspiration—inhale[63]— which is poetic breathing.

It can be said that poetry, in the generic sense of an artistic attitude, is a secondary matrix of our techno-scientific societies that expresses its animistic part. Trying the posture of the poet is, in a way, "*putting spirits back into the forest as François Terrasson advocated, i.e., reinjecting the symbol, the link of reciprocity and friendship*" (Planche, 2017) in the image of the animist who recognises an animated and talkative Nature, capable of informing him in reciprocity.[64]From a pedagogical point of view, we can see that the experience in nature makes the third party with the idea of good writing for the children in difficulty.... but also for the adults embedded in their pre-conceptions. The productions are much more personal! The environment will give them a hand to write, to find themselves rather than to find the performance, to let themselves write rather than to think about writing well.

In the manner of the Inuit, who want to listen to their intuition while distrusting categories and schedules, ESD, through sensitivity, invites, under the guise of the artist's posture, to stop the scholarly gaze in order to encounter the environment in a different way, in a more direct and sensitive manner, leaving aside the categories of thinking that sclerotize the perception of reality. As Kirby reminds us, "*Human beings would be as if trapped in a network of discursive interpretations that would make any direct experience and knowledge of nature impossible* " (2011). We can highlight the points in common between the poet who broadens his vision by leaving the blinders of cladism, the animist who multiplies perspectives on the world by recognising that each element of Nature has a different point of view to put forward, and the ethologist of the twenty-first century, which highlights the multiplicity of animal perceptions of reality, including, for example, the pointillist point of view of the frog (Clément, 1999), thus rebonding off with a nod to the artist!

Nature, as an ecosystemic symphony, allows us to make Art, just as an open and contemplative artistic attitude allows us to see Nature penetrate its interstices. The environment can be a lever for bringing perceptions and ideas into the world when, by listening to the wind, "*a whispering of the stones, a rustling,*" says Malaurie, the artist and the Inuit place themselves in symbiosis with Nature. But also, Art launches tentacles of colours and forms to establish another link with the environment by shifting the borders to highlight different entanglements between the things of the world. Art captures the invisible between the lines and brings to light different contours of reality to open up to new dimensions and to the threshold of other paradigms and understandings. Picasso explains, "*Painting is never prose, it is poetry, written in verse with plastic rhymes [...]. Plastic rhymes are forms that resonate with each other, respond to other forms, or to the space that surrounds them*" (Gilot, 1965), at the image of the shaman who speaks to us of an invisible reality, beyond, through or beyond, Art and shamanism, surfing between the lines of reality in a form of common nomadic thought?

In conclusion

Thus, education through the sensitive, by reinjecting the subject, proposes a rehumidified education against the single and dry overhanging, though not identified as such, because it is not conscious of its imperialisms. In order to make a leap towards another paradigm, it is necessary to enter the century of the circle of points of view against the universe of the functional and very poor single thought (without vibration of the living and depriving itself of the other points of view: that of the educated learner being subject to univocal standards, that of the human plural and that of the non-human living). With the chthulucene, we are suggested to let tentacles, hypotheses, poems hang out, to probe perhaps, in my opinion, what lies behind, underneath, and beyond time and space?

While modern societies have an impoverished vision, as we impoverish the soil. In the ecosystemic sense, nature is basically the same as the concept of *Pacha-mama*: under the earth, which is teeming with worms, there is swarming life. The earth is animated by a thousand lives and a thousand living beings, by thousands of paradigms and points of view on the world. This is rich and fascinating, in contrast to the rationalisation paradigm, which invites us to trample, devitalize, and flatten all this life. It seems necessary today to reinject points of view: that of wild thought and of each society, that of each living being (animal and plant points of view) and of each individual (experience of the subject in education). The earth has fallen silent; we must be silent (or rather, silent about the Occidentalocene and human categorical thinking) in order to re-establish links and subjects. Let's become sensiblocene (Planche, 2022) to reverse the Anthropocene! It is not a question of cultivating an artificial or anorexic happy sobriety by thinking of the paradigm reversal as a loss of privileges compensated by virtue, but rather of valuing the virtuous circles of the living (those of nature and physis, but also when art mobilises the living within ourself and when ethnography reanimates the breath of the ancestors) by thinking of the reversal of values in terms of well-being with oneself and with the planet. For what morality seeks to build bridges with its own entrenchments?[65]

While ethnology allows us to apprehend "wild thought", even to experience it through the channels of ethnopedagogy, art allows the Western world[66], in its imbalance, to reinject this form of thought connected by forms, colours, sounds, and is just as valid for deciphering the world. Otherwise, we run the risk of remaining lame of a vital dimension or of remaining aseptic in an anthropocene civilisation that distances itself from the language of the earth and of the Earth, from the language of birds, from the language of the Other of the world. However, it is essential today to undo the cascading overhangs that interlock, to forge a way of thinking in which there is no longer a centre[67]: the overhang of the human over the animal and the earthly condition, the overhang of the white man over the people who are the guarantors of savage thought, the overhang of the *masculine* over the *feminine*[68], the teacher over the student. Knowing and approaching the Other of the world, this face hidden from Western thought or this blind spot of the living and of science is necessary for an education in terrestrial humanity resituated in the network of the living[69].

In order to curb the ecological crisis, is it a matter of re-establishing this living and the link to the living in order to turn the Anthropocene around? The living at several scales and levels: the living of the fructification of art and symbol in the face of a disenchanted reality, the living of the experiences[70] of the ancients in the face of a devitalized relationship to the living environment, the link between relocated living people in order to co-construct a living relationship to the territory (Faburel G. and Vialan D., 2018) and the link to other living beings on Earth in the face of a relationship with non-humans and the soil that is sanitised and mortifying. Above all, is the ecological crisis a crisis of the link, of the vibrant and of vitality? Art and ethnology (anthropology, ethnopedagogy, and ethnography) will re-entangle all these sensitive and connecting threads, with a nod to chthulucene, to re-establish the bridges of meaning and restore consistency, volume and "full" to our relationship with the world in order to fight against this aplatocene civilisation, subjectmortifocene. This education through the sensitive, by re-creating bridges with oneself, with the ground and the symbol, participates by re-injecting the cells of life as a resilience to the Anthropocene and the domination of the artefact in our lives.

Endnotes

[1] "There is no sustainability without mutual understanding." Jan Činčera, 11th World Environmental Education Congress Building Bridges in Times of Climate Urgency. 14/18 mars 2022.

[2] http://science-et-art.com; https://www.arts-ere.net/arts-plastiques; https://www.arts-ere.net/arts-plastique---sea

[3] Cf. D. B Zandvliet (2019).

[4] "Rationalism" and "positivism" in the structural sense: these "isms" are not to be confused with the necessary rational.

[5] http://science-et-art.com/Projet-Regards-de-Rhone-action

[6] http://science-et-art.com/Liens

[7] The SeA teaching tools that I designed and some of them illustrated (drawings, watercolours and poems) are gathered in the kit http://science-et-art.com/Mallette-Pedagogique, sponsored by Edgar Morin and Jean-Marie Pelt (preface to the booklet "Balade poétique", 2008), by Philippe Meirieu, by Jean Malaurie and by François Terrasson, with the praise of André Giordan and Lucie Sauvé. Several of these tools are included in my book after a process of decontextualisation, and the book includes additional workshops. http://science-et-art.com/Les-Balades-Science-et-Art; http://science-et-art.com/Les-Ateliers

[8] The evaluation shows that this project has fostered the creativity of the pupils, their families and the teachers, and that it has developed an interest in their territory. https://augustinbouquetin.jimdofree.com/travaux-des-%C3%A9coles/

[9] https://augustinbouquetin.jimdofree.com/balade-familiale/

[10] "Nature, Culture, Science, Art" are capitalized and placed in the singular, as concepts, to include all the cultural representations these notions convey.

[11] Design example: http://science-et-art.com/Resultats-en-images-d-une-enquete

[12] See speech E. Planche https://centrere.uqam.ca/2021/jour-de-la-terre-arts-et-ere-lancement-conference-et-table-ronde-22-avril-2021/ https://www.youtube.com/watch?v=KfCM_jjJKHg (at 18 min).

[13] The President of the association SeA, Science et Art, Philippe Louisgrand, artist and honorary director of the International School of Art and Design of St. Etienne http://www.philippelouisgrand.fr/ and his vice-president, François Laplantine, author of more than 30 books, professor emeritus of ethnology who contributed to the creation of the department of anthropology at the University of Lyon2, played the role of expert in these disciplines, i.e., art and ethnology. https://fr.wikipedia.org/wiki/Fran%C3%A7ois_Laplantine

[14] My work is in line with the CIRET charter https://ciret.hypotheses.org/sites-des-membres-et-contacts, which defines transdisciplinarity as a vision "resolutely open insofar as it goes beyond the realm of the exact sciences by their dialogue and reconciliation not only with the humanities but also with art, literature, poetry and inner experience," carrying the project of an "authentic education [which] cannot privilege abstraction in knowledge

[ ] must teach to contextualise, concretise and globalise" and must "reassess the role of intuition, imagination, sensitivity and the body in the transmission of knowledge".

[15] Cf. E. Planche, publication of three watercolours and poems in tribute to Basarab Nicolescu: Omul cosmodern / L'Homme cosmoderne / The Cosmodern Human, Iaşi: Junimea, 2022.

[16] Jean Malaurie has given me a platform to imagine a teaching that re-inspires empowerment in the people of Northern Arctic Siberia: cf. "Réveiller la force, Cinq leçons pour ré-animer les peuples du Nord", in J. Malaurie (2019). The title "Reveiller la force" reminds us that it is indeed a question of loss of confidence, of loss of an "imaginary that becomes westernised" and collides with the flat screens of an uprooted thought without resonance and fructification of the myth. It is a question of reawakening strength, which is to say, of reinjecting vitality through Art and re-anchoring ourselves to the ground as levers of hope and desire to learn.

[17] Adding, however, an "art and science" dynamic. See E. Planche (1996): *An Ethnographic Report in Poetic Form.*

[18] For a distinction between ethnography and ethnopedagogy, see p. 40 and 41. E. Planche/Collective Writing (2011), in *Regards d'ici et d'ailleurs.*

[19]Excerpt from a summary of B. Jickling's speech of Tuesday 15.03.2022 "Reflections on arts, ethics, and environmental education", *SessionT02—Arts, Ethics, and Environmental Education*, 11th World Environmental Education Congress Building Bridges in Times of Climate Urgency. 14/18 mars 2022.

[20] "*[Rational thought] is determined by the intelligible: we evacuate our sensations to manipulate concepts. Conversely, wild thought calculates, not with abstract data, but with the teaching of sensitive experience: smells, textures, colours. In both cases, man is trying to decipher the universe, and savage thought, in its own way, succeeds as well as modern thought",* C. Lévi Strauss reminds us (quoted by G. Sorman, 1989).

[21] It is indeed a question here of the idea of art and the idea of science.

[22] "*An anthropological aesthetic leads to freedom from a certain number of oppositions, and in particular from the sterile dichotomy of a rationalocentrism without morality and a moralising humanism. It allows us to stop opposing 'science', born of the disappointment of illusions and leading to a certain submission, and 'art', which would be seduction, delectation, relaxation, entertainment, (...) oblivion of oneself and of others.*" (Laplantine, 2005, p. 213).

[23] http://science-et-art.com/Parrainages

[24] While the SHS attempts to stand back by presenting abstract individuals to bring out ideal types, literature probes the sensitive digressions of human reality that can substantiate the notion of the subject.

[25] M. Lussault évoque la notion d'urbanocène https://ecoleanthropocene.universite-lyon.fr/

[26] See M. Tadié, in J. Malaurie, (2002).

[27] Without forgetting that the landscape is everywhere and constantly shaped by human activities (cf. F. Braudel, 1995). It is instead a question of dosage because the Anthropocene is also an urbanocene, meaning an excess of the anthropisation of spaces.

[28] Caged calves, hens and rabbits, and castrated pigs are extreme manifestations of the negation of the living subject from a rationalisation perspective.

[29] The exacerbated illustration of this "objetisation"—making the animal an object and no longer a subject—is the treatment of animals in industrial breeding.

[30] For the notion of *Pacha-mama*, see J. Malaurie (2018), pp. 155-156.

[31] and thus become secondary and non-heuristic values or at least framed in culture, separate from the sciences.

[32] I. Stengers commenting on D. Haraway tells us that we are in an era where we tell stories of connections and ramifications. https://www.youtube.com/watch?v=e-C990OWj0Q_

[33] SFR = Services Rendered by Nature; SFN = Services Based on Nature.

[34] http://science-et-art.com/Balades-citadines-grand-public, and E. Planche (2018), p. 220/236, 394.

[35]Extract summary presentation 15.03.2022. S. Blenkinsop (Auteurs B. Jickling, M. Morse) " Wild pedagogies and the educational imagination", *SessionT08—Environmental Education in the Anthropocene,* 11th World Environmental Education Congress Building Bridges in Times of Climate Urgency; 14/18 mars 2022.

[36] Even so, in order to avoid the risk of unlimited solipsism, all this sensitive material must be regulated by the collective and the debate, and/or organised in sequences of alternating approaches.

[37] Other workshops on the theme will mobilize the sensory and kinesthetic with the silent crossing of the bridge (E. Planche (2018), p. 252) or observe its users for a socio-artistic investigation. Cf. *The Rhône, the tool of the wonders of the river*, SeA, Stepping over the river, 2008. http://science-et-art.com/Les-Ateliers

[38] E. Planche (2018), p. 256-257.

[39] E. Planche (2018), p. 282.

[40] E. Planche (2018), p. 350.

[41] E. Planche (2018), p. 307.

[42] The preoccupation with the bridge that cuts the current (as shown by the ritual of throwing mannequins into the Tiber to gain clemency for this offence) refers to the symbolic significance of the crossing: passage to an Other

and an Elsewhere: another shore, another world, another paradigm…

[43] "Art crosses things, it goes beyond the real as well as the imaginary" says Paul Klee (Quoted by JP Duvaleix, 2005).

[44] E. Planche (2018), p. 292.

[45] E. Planche (2018), p. 299.

[46] E. Planche (2018), p. 405.

[47] E. Planche (2018), p. 396.

[48] E. Planche (2018), p. 400, 401.

[49] E. Planche (2018), p. 415.

[50] E. Planche (2018), p. 420.

[51] Extract presentation session of Tuesday 15.03.2022, *T02- Arts, ethics and environmental education.* Chairs: B. Jickling , J. Dlouha , L. Henderson, 11th World Congress on Environmental Education. Building bridges in times of climate emergency, 14/18 March 2022.

[52] Wind with the notion of air is central in Indigenous North American and Indian traditions with the breath (D.Abram, 2021, p.42 and chapter 7; F.Varela, 2017).

[53] Either privileged thought among non-naturalistic peoples* (*in the sense of P. Descola, 2005, p. 323).

[54] Understood as a mode of thought different from modern thought because C. Lévi-Strauss adds that "In both cases, man endeavours to decipher the Universe, and savage thought, in its own way, succeeds as well as modern thought. (G. Sorman, 1989).

[55] See diagram, Edith Planche *'The Western model and the animist model in a generic sense'* (Planche, 2018, p. 87) and http://www.science-et-art.com/IMG/pdf/ENS_Cachan-2.pdf p. 5 taken from E. Planche, (2011), p. 217.

[56] as *raison d'être* of the ecosystem.

[57] This is the approach of the Institute for Environmental Learning https://www.eco-learning.org/: "*Now, in the early 21st Century, we have the opportunity to draw on even older discourses, such as the land-based knowing of Indigenous peoples around the planet, to reimagine how we conceive of our relationship to and responsibility towards the interconnected network of eco-systems that make up our home. As the Institute for Environmental Learning moves into its next decade, we seek to build relationships with Indigenous community members and Knowledge Keepers to renew our vision of land-as-teacher, to move away from ideas of land as commodity, as recreational, and consumable.*" Extract *Director's message* David Zandvliet (SFU) and Shannon Leddy (UBC). L. Sauvé also mentions "*the opening up to Amerindian or Oriental cultures*", which will challenge the very notion of development (Sauvé, 1999, p. 17).

[58] The elf, an anthropomorphic figure, represents this access to the unknown space of Nature with a sensory perceptive entry. Big ears to hear, a long nose to smell. In Burkina, the spirit of Nature, the Kinkirga, is immediately available in the landscape (in children's drawings), whereas in the West, we use poetic metaphor. https://irp.cdnwebsite.com/93da38ae/files/uploaded/N%C2%B04_LES%204%20BALADES%20SENSIBLES.p df) p. 78-80.

[59] C.f. also the concept of eco-training: G. Pineau (1992).

[60] Cf. also the phenomenology https://fqs.hypotheses.org/sensibilites

[61] Cf. France Culture: "Le Facteur Cheval: que celui qui n'a jamais rêvé lui jeter la première pierre..."-*Le Palais idéal du Facteur Cheval* ©Radio France by S. Nauleau & C. Diger, 22.03.2012. *The Ideal Palace*, the work of Joseph Ferdinand Cheval, known as *Le Facteur Cheval*, was classified as a historical monument in 1969 by André Malraux. His work started by "chance" during his postman's round with the stone he stumbled on, just like our poetic walks, which invite us to let ourselves be carried by the movement of our encounter with nature: "*I remain attentive to my sensations, to the tracks that open up. I bump into a small pebble, I pick it up, it has pink reflections*", cf. Planche (2018), p. 144 and workshops, p. 144 et ateliers p. 220/236 ; 246/253 ; p 265 ; p. 266 ; p 368/378.

[62] Maurice Leenhardt reports that a Canaque told him that "it was Christianity that had taught him that he had a body. That is to say, first of all, an object that was separate from the soul, instead of engulfing the spirit in matter —a body, that is to say, also an object separate from the other objects of nature, from the totemic stones, from the animals and plants, as well as from the name one gives it, from all its belongings." (Bastide, 1965/1993)

[63] To breathe is to inhale and exhale, "It is to be more present to the Life that circulates, it is to enter into the movement of Life" (Xerri, 2020).

[64] Reciprocity is nevertheless friendly and/or conflicting.

[65] This unconsciousness behind the self-righteous word reproducing imperialisms: the fault is always with the other.

[66] In other words, to reinject wild thought profanely and provide exoteric answers to the following question: "Do indigenous knowledges offer pedagogies of kinship with the land and ancestral relationships with the entities

of nature? And how can they be used appropriately? Extract from the presentation of the session of Tuesday 15 March 2022. *T08 -Environmental Education in the Anthropocene*, Chairs: M. Paulsen, S. Blenkinsop, K. Malone, 11th World Environmental Education Congress, Building bridges 14/18 mars 2022. [67] Even though Tim Ingold points out that the term anthropocentrism is inappropriate (Rigal, 2022) for an overhanging (pyramidal, as I have represented it cf. E. Planche, 2018, p 87 or http://www.science-et- art.com/IMG/pdf/ENS_Cachan-2.pdf, p. 5. (repeated in E. Planche (2011), p. 217)

[68] with associated representations and for 'masculine' and 'feminine' poles reconciled in each person, woman and man (E. Planche, May 2022, p. 10).

[69] We can put together here all the pieces of the chthulucene's attempt, which has the merit of building bridges into other places and Others that sometimes remain blurred and undefined: connections between the living, the capacity to choose and to be responsible, which would pass through the capacity to feel and to 'keep the heart', links between science and the arts, the real and the imaginary, subjects and objects, but also consideration of animism as something to be conjugated in the past-present-future...

[70] The rural farmer of the 19th and early 20th centuries established a dialogue and a survival link with nearby nature, offering compositional models that are re-visited in the 21st century for a "responsible" sharing of nature–culture spaces.

Reference

Abram, D. (2021). *Comment la terre s'est tue*. Pour une écologie des sens. Traduit par D. Demorcy et I. Stengers. Titre original *The Spell of the Sensuous* (1996). La découverte poche, p.42 et chapitre 7.

Asselin, H. (2019). « Pour en lire plus : Éduquer à l'environnement par l'approche sensible. » In *Volume 15 – 1 | 2019, L'éducation à l'environnement au sein des aires protégées et des musées Recensions d'écrits*, Education Relative à l'Environnement. Recherche, regards, réflexion. https://doi.org/10.4000/ere.4545

Bachelard, G. (1983 [1934]). « L'épistémologie non-cartésienne », *Le nouvel esprit scientifique*, Paris, PUF (Quadrige), p. 139-183. Dans la suite, cet ouvrage sera abrégé NES.

Banywesize, E. M. (2007). « Edgar Morin et le réenchantement des sociétés humaines », *Sociétés 2007/4* (n° 98), p. 23-39, .p.31.

Bastide, R. (1965). « La pensée obscure et confuse ». in *Le Monde chrétien*, n° 75-76, juillet-décembre 1965, p. 137-156. and *Bastidiana*, n° 2, avril-juin 1993, p. 77-89.

Braudel, F. (1995). *L'identité de la France.* Tome 1, Espace et Histoire, Arthaud.

Clément, P. (1999). « Autant d'espèces, autant de mondes ! Des mondes animaux ». *Science et Avenir*, 119 (7-8), p. 20-25.

Descola P. (2010). *Diversité des natures, diversité des cultures*, Les petites conférences, Bayard, p.69-70 Descola, P. (2008). « A qui appartient la nature ? », in *Dossier Les défis de l'écologie*, 21 janvier 2008. Descola P (2005). *Par-delà nature et culture.* Paris : Gallimard.

Devereux, G. (2010). *Essais d'ethnopsychiatrie générale*, Gallimard. Duvaleix, J.-P. (2005) *Journal des peintres*, www.journaldes peintres.com.

Faburel G. & Vialan D. (2018). « Chapitre 10 - L'engagement par la sobriété choisie : vers une construction (cosmo) politique de la transition énergétique ? Le rôle du bien-être », pp. 173-186, In Format OpenEdition APA MLA, Villalba B (dir.) ; Semal (dir.) *Sobriété* énergétique *: contraintes matérielles,* équité *sociale et perspectives institutionnelles*, Versailles, Editions QUAE. https://books.openedition.org/quae/16567?lang=fr Gilot, F. et Lake, C. (1965), *Vivre avec Picasso*, Calmann-Lévy, p. 146.

Haldas, G. (1977). *L'État de poésie: carnets 1973. Les Minutes heureuses.* Éditions L'Age d'homme. Haraway, D.-J. (2016). *Staying With the Trouble: Making Kin in the Chthulucene*, Experimental Futures. Jickling, B. ; Blenkinsop, S. ; Timmerman, N. ; De Danann Sitka-Sage, M. (2018). *Wild Pedagogies: Touchstones for Re-Negotiating Education and the Environment in the Anthropocene*, Palgrave Studies in Educational Futures.

Kirbi, V. (2011). *Quantum Anthropologies: Life at Large.* Durham, North Carolina: Duke University Press. Laplantine, F. (1996). La description ethnographique. Paris: Nathan, p. 75.

Laplantine, F. (2005). *Le social et le sensible : Introduction* à *une anthropologie modale.* Paris : Téraèdre. Laplantine, F. (2018). *Penser le sensible*, Agora.

Latour, B. (2018). *Face à Gaia. Huit conférences sur le Nouveau Régime Climatique.* Paris : La Découverte. Malaurie, J. (2002). Pas à pas avec vers les Inuit. Dans *Séminaire de Jean Malaurie 2000-2001, De la vérité en Ethnologie.* (p. 107-148). Paris: Economica.

Malaurie, J. (2015). *Lettre* à *un Inuit de 2022*, p.108 109 Malaurie, J. (2018) *Oser, résister*, CNRS Editions.

Malaurie J. (2019). (Dir.) *Arctica,* œuvres *2, Tchoukotka 1990. De Lénine* à *la Pérestroïka*, CNRS Editions. Morin

E, (1984). *Sociologie*, Fayard.

Morin E., Motta, R. et Ciurana, É.-R. (2003). Éduquer *pour l'ère planétaire : la pensée complexe comme méthode d'apprentissage dans l'erreur et l'incertitude humaines.* Paris: Balland.

Morin E., (1999), *La tête bien faite, repenser la réforme, réformer la pensée*, Seuil.

Morin E., (2000), *Relier les connaissances. Le défi du XXIe siècle,* Seuil. « C'est précisément en permettant la connexion entre ces deux pôles, le scientifique et le littéraire, que l'on peut construire une culture commune. Inventer de nouvelles humanités ».

Morin E., (2014), *Enseigner* à *vivre, Manifeste pour changer l'éducation*, Arles, Actes sud, Playbac. O'Dy S., (1993, 23/09) L'affectivité nous façonne, *interview de Boris Cyrulnik.* L'express.

Pardo T (2002). *Héritages buissonniers.* Éléments *d'ethnopédagogie pour l'éducation relative* à *l'environnement.* La Caunette: Éditions Babio.

Pineau G. (1992) *De l'air ! Essai sur l'écoformation.* Paris: L'Harmattan.

Planche, E. (1989, 1999) Chapitres 5, 9 (E. Planche) et chapitre 15 (en collaboration avec J.F. Vincent), in *Pays de Thiers, le Regard et la mémoire,* sous la direction de D.Hadjadj, Institut d'Etudes du Massif Central, Editions CREER, réédition Parc Livradois-Forez.

Planche E, (1996) « Quand l'ethnologie se donne à voir et à dire, aspects, formes et enjeux de la média(tisa) tion », Actes des 11e rencontres de l'Association Rhône-Alpes d'Anthropologie, Charlieu.

Planche, E. et Mignard C. (photographe) (1996), Jour de pêche en Dombes, Editions Aléas.

Planche E. (2007), « Science et Art ou Créer pour aimer et apprendre », Dossier « Artistes et ethnologues, enjeux de la création » in *Revue de l'Association Rhône-Alpes d'Anthropologie*, pp. 55 à 61.

Planche, E. (2008). « Balade poétique ». In *Le Rhône, la mallette des merveilles du fleuve, outil magique et pédagogique.* La Mulatière : SeA, Science et Art.

Planche, E (2011) Co-auteure écriture collective Partie 1 « Une opposition Nature/Culture à dépasser », pp. 10-16 ; pp. 38-41, in *Regards d'ici et d'ailleurs, Quand l'éducation à l'environnement explore les liens que tout peuple tisse avec la nature*, Réseau Ecole et Nature.

Planche, E (2011) « La balade poétique ou la posture de l'animiste », pp. 64-65, in *Regards d'ici et d'ailleurs, Quand l'éducation à l'environnement explore les liens que tout peuple tisse avec la nature*, Réseau Ecole et Nature.

Planche, E (2011), "Le rapport de l'Homme à son environnement et la notion de sujet," in *L'idée de nature dans l'éducation et la médiation scientifique*, ARTheque - STEF - ENS Cachan, pp. 213-226.

Planche, E (2017), « Créer pour recréer le lien avec l'environnement », In Volume 14-1 *Arts et* Éducation *relative* à *l'environnement*, Éducation relative à l'environnement. Regards. Recherches. Réflexions, Sous la direction de Vincent Bouchard-Valentine et Lucie Sauvé. https://journals.openedition.org/ere/2674

Planche E. (2018) *Eduquer* à *l'environnement par l'approche sensible. Art, ethnologie et* écologie, Chronique Sociale. PRIX DE LA RECIPROCITE FORESCO 2022 (Mouvement français des Réseaux d'Échanges Réciproques de Savoirs®)

Planche E. (2019) sous la Direction de Jean Malaurie, « Réveiller la force, cinq leçons pour réanimiser les peuples du Nord », *Arctica,* œuvres *2*, pp. 679-697.

Planche, E. (2022) « Book Review, To educate about the environment through a sensitive approach. Art, ethnology and ecology. Chronique Sociale, June 2018», in *Environmental and sustainability education in compulsory education: challenges and practices in Francophone Europe*, Environmental Education Research. https://www.tandfonline.com/doi/full/10.1080/13504622.2022.2146657 https://www.tandfonline.com/eprint/UKQTQTZTPXEUPVTPPFVM/full?target=10.1080/13504622.2022.21466 57

Planche, E. (2022). « Nous sommes coupés du sol et du symbole ! », Revue annuelle, Mai, *Artistes et animateurs, ensemble pour créer l'enchantement*, Graine Nouvelle-Aquitaine, Plumes d'Orfée.

Platon, Œuvres *complètes, tome 4, 3ème partie, Phèdre*, texte établi et traduit par Léon Robin (1970). Paris : Les Belles Lettres, p. 32-33.

Rigal, A. (2022, 21/12). Repenser l'évolution, la Terre et la science. Entretien avec Tim Ingold. Anthropocène 2050. *Ecole Urbaine de Lyon.*

Rimbaud, Arthur - Edité et commenté par Schaeffer, G *(1975), Lettres du voyant. (13 et 15 Mai 1871).* éditions Droz.

Sauvé, L. (1999) « Environmental education, between modernity and postmodernity – Searching for an integrative framework. ». In A. Jarnet, Jickling, B., L. Sauvé, Arjen Wals et Priscilla Clarkin (dir.). (2000). The Future of Environmental Education in a Postmodern World? Whitehorse: *Canadian Journal of Environmental Education.* Vol. 4, p. 9-35.

Sorman, G. (1989). *Les vrais penseurs de notre temps.* Paris : Fayard.

Tadié, M. (2002). Perceptions extra-sensorielles des populations primitives. Dans Malaurie, J., Séminaire *de Jean Malaurie 2000-2001, De la vérité en Ethnologie,* Paris : Economica, p. 31-53.

Terrasson, F. (1997) *La peur de la nature.* Paris : Sang de la Terre.

Thésée G. et Carr P. R. (2008a) « Une proposition d'élargissement de la dimension critique en éducation relative à l'environnement : la résistance éco-épistémologique ». *Education Relative à l'Environnement* 7. http://journals.openedition.org/ere/3193.

Varela F, Thompson E, Rosch E (2017). *L'inscription corporelle de l'esprit-* Traduit par V. Havelange. Sciences cognitives et expérience humaine – Points. Essais.

Varela, F. (2017) *Le cercle créateur - Ecrits (1976-2001),* La couleur des idées, Seuil.

Vincent G, (1992) « Patrimoines en débat, Construction de Mémoire et Valorisation du Symbolique », in *Les Papiers, Economie-Société,* Communication, Presses Universitaires du Mirail, N°9 Printemps 1992, p. 133-145, à propos du rapport E. Planche et M.-J. Peltier, « Recherche sur le patrimoine industriel textile en roannais », réalisée en réponse à l'appel d'offre du PPSH *Pratiques culturelles et institutions de conservation du patrimoine dans la Région Rhône-Alpes,* sous la direction de Guy Vincent, 1991.

Xerri, J.-G. (2021). *La vie profonde.* Editions du Cerf.

Zandvliet, D.-B. (2019) *Culture and Environment: Weaving New Connections,* Brill.

Chapter 5

BioLearn

International project about biomimicry in education for sustainability

Éva Neumayer[1], Márta Kurucz[2], Milada Dobiasova[3], Klaudia Medalova[4], and Richard Medal[5]

Introduction

BioLearn (Learning from Nature - The biomimicry in Education) project was supported by an Erasmus+ strategic partnership, started in September 2018 and ended in October 2021. Partners of the project: Magosfa Foundation (Hungary); SEVER (Czech Republic); CEA (Slovakia); Biomimicry NL (the Netherlands); Wild Awake (United Kingdom); and the Center for Learning with Nature (U.S.A.).

Biomimicry is a design method that uses knowledge from nature as inspiration for sustainable design. It is an interdisciplinary approach that brings together nature, biology, design and technology. Similar expressions are bio-inspired, biomimetics, and bionics.

Keywords: biomimicry, sustainability, learning from nature, educational modules

Research questions/Purpose of the program

The aim was to elaborate educational modules about biomimicry for students aged 11–16 and their teachers and introduce them to biomimicry thinking—a way that can help make our world more sustainable. The modules are available on the project website (www.biolearn.

[1] Magosfa Alapítvány, Magyar Környezeti Nevelési Egyesület

[2] Magosfa Alapítvány; Magyar Környezeti Nevelési Egyesület

[3] Středisko ekologické výchovy SEVER

[4] Centrum Environmentálnych Aktivít

[5] Centrum Environmentálnych Aktivít

eu) in five languages: Czech, English, Dutch, Hungarian, and Slovakian.

A part of the project was to elaborate and try out teacher training programs in the European countries of the project. Within the project, we also organized a study trip to the Netherlands in the field of biomimicry, with the participation of 55 teachers or educators from other European partner countries (Czech Republic, Hungary, Slovak, and the United Kingdom). At the end of the project, national conferences were organized.

Methodology/Program Description

To understand **biomimicry** and its importance in sustainability, it is helpful to recall some **examples** and stories.

Probably the most well-known example is Velcro[1]. In 1941, George de Mestral, a Swiss engineer, was walking his dog and noticed it was hard to pull out burdock fruits from the fur of his dog (and from his own clothes). He was thinking of a new and strong fastener anyhow and started the development based on this idea. It was not an easy and quick process, but by the early 1960s, he invented Velcro, which was first used in astronauts' clothing. Our life is almost unimaginable without this technology.

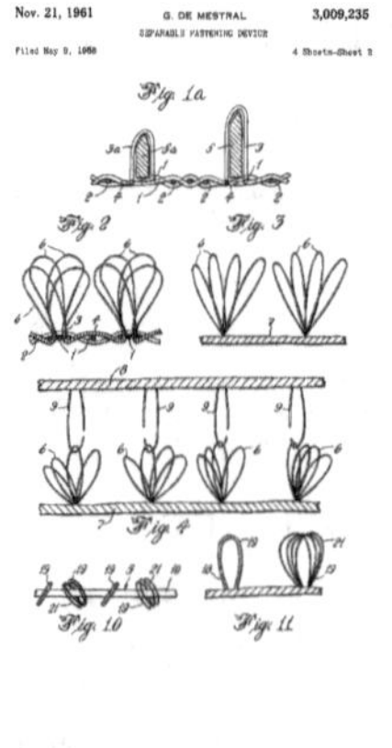

Figure 1: Patent of Velcro.

Warrior Goddess @ Ellyn Lyle 2019

The other well-known example is the Shinkansen, the high-speed train in Japan. Japanese engineers developed the fastest train in the world in the 1960s. It was very fast, but coming out of a tunnel caused a big noisy "explosion" as the train pushed air out of the tunnel. Eiji Nakatsu, the head engineer, was a birdwatcher and knew the fishing ability of kingfishers. Eiji predicted that as he and his colleagues tried out different-shaped models for the train, the model most like a

Figure 3: Kingfisher arrives at the water

Figure 4: Different Shinkansen shapes

kingfisher's beak would be the quietest. After the train's front end was redesigned with a shape much like a kingfisher's beak, it proved to be much quieter and used less energy to travel.

A good example of the sustainability usage of biomimicry is how mushroom mycelium can be used as packaging and insulation to replace Styrofoam. In the 1990s, two young men in the U.S.A. (Gavin McIntyre and Eben Bayer) invented this method by growing myceli-

Figure 5: Mycelium packaging

um in certain shapes and drying them out. Their idea became a bigger enterprise (Ecovative LLC—The Mycelium Technology Company https://www.ecovative.com/), and it is used in many countries for different purposes.

The fourth example is about mimicking the microstructure of certain organisms. Researchers examined the surface of lotus leaves, searching for the answer as to how the plant keeps its leaves clean. They found that the surface of the leaf is not flat but is covered with a kind of micro-fuzz. This fuzz causes the drops of the rain to run off, taking the pollution with them. Mimicking this structure, self-cleaning textiles

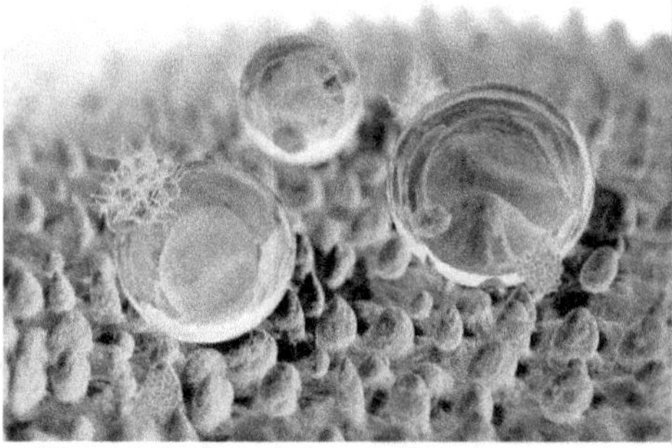

Figure 6: Surface of a lotus leaf.

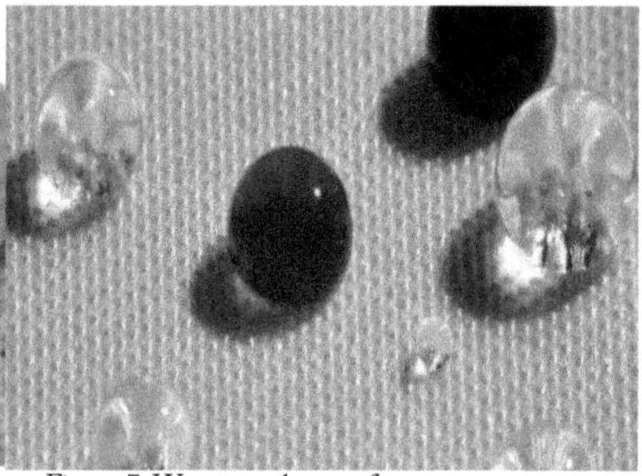

Figure 7: Water-repelent surface

and paints were invented.

The name "biomimicry" and the use of the ideas of nature in sustainability comes from J. M. Benyus and her book.[2]. She began to think in the 1980s that nature had solved most of its problems during the last 3.8 billion years, and still worked in a sustainable way. She believed we could learn from nature and started to link biologists and engineers. With early successes, biomimicry became a kind of science.

She continues her work at the Biomimicry Institute (https://biomimicry.org/), and thousands of great biomimicry examples are available on their website (https://asknature.org/).

Biomimicry has several **principles**. Depending on the author, these can be divided into different categories, and the numbers are varied (5, 6, 9, 12). In our work, we used the original nine principles according to J.M. Benyus. These principles fit with sustainability and describe how nature operates:

1. Nature runs on sunlight.

   Nature uses sunlight as the main source of energy. Organisms use heat and UV radiation from this never-ending source. The main base of life is photosynthesis.

2. Nature uses only the energy it needs.

   Animals only take the nutrients they need; plants do not absorb more water than is necessary. There is storage in nature as well, but not in that enormous amount as in the world of humans.

3. Nature fits form to function.

   A tree is rooted in the ground to draw water and nutrients from the soil; it spreads its branches and leaves wide to increase surface area and absorb the sunlight to produce energy and grow. Nature creates designs for the function they provide.

4. Nature recycles everything

   There is no 'away' to throw things. Everything produced in nature is biodegradable; there is no waste. What is waste in one process becomes raw material in another one.

5. Nature rewards cooperation.

   We see competition in nature, but only when it's impossible to avoid; in general, competition costs too much energy. On the other hand, very little in nature exists in isolation.

6. Nature banks on diversity.

   Diversity is one of nature's best insurance policies. When one food source is unavailable, others can be found. Plants use several different strategies to spread seeds or defend against predators. We know that species with limited genetic diversity have more difficulty adapting to environmental change, and that ecosystems rich with diversity are more stable.

7. Nature demands local expertise.

   Nature's systems are inherently local. Certain species thrive under specific conditions; local and regional weather patterns matter, as do other conditions such as soil, air quality and water temperature.

Relationships are created locally, and local resources are used.

8. Nature seeks balance.
   Ecosystems will always try to keep in balance. More mice? Then, you will see more owls to feed on the mice and keep the population in balance. Forest fires are a great example of a natural phenomenon that renews and refreshes, reducing excessive growth and allowing for regeneration. Every natural system has a tipping point, a carrying capacity or a state of disequilibrium that triggers a change to a different state.

9. Nature taps the power of limits.
   Unlimited growth on a finite earth is not a good idea. All living things are governed by limitations; age, climate, population density, and many other factors determine how species and systems develop. Nature has found ingenious ways to work within these limits to be as productive as possible over the long run.

The **biomimicry design process** can be made in two ways: The first way starts with observing a natural object and continues with the question: 'What can we learn from this?'. Actually, we can learn numerous things from any organism we examine. After we identify and define certain features and functions of an organism, we can brainstorm about what can be useful in our lives and invent something based on the ideas. Then, we should match it with biomimicry principles to check if the idea is sustainable and fits nature's operation. In that case, where everything seems to be alright, the emulation continues.

The second way starts with a given challenge, and the question is, 'How would nature solve this challenge?' In this case, we search for solutions in nature that can help in that problem or challenge.

The key thing is, in both cases, the translation from technical or everyday language to biology. We can easily see this in one of the examples above, about the surface of lotus leaves. When our clothes are dirty, we often think that the problem is the quality of the detergent, so we focus on improving the detergent. It is hard to find detergents in nature (however, we can find plants containing saponins), but our real question is not about the detergents but about the cleanness of clothes. We can ask: How does nature keep surfaces clean? We will

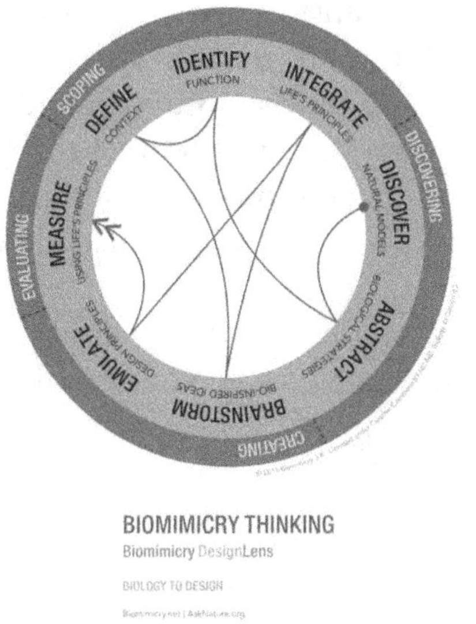

Figure 8: Biology-to-design design process model.

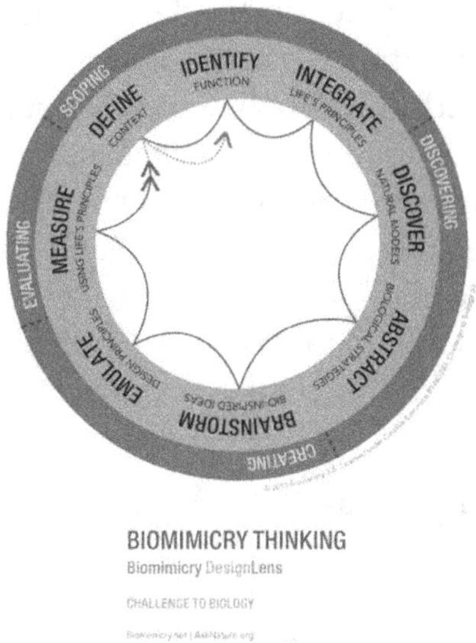

Figure 9: Challenge-to-biology design process model.

find numerous solutions for this problem in nature—one is the lotus leaf, which we can observe closely and how it works, and we can obtain ideas from it.

Our project was about biomimicry and education, and our closer aim was to promote the usage of biomimicry in schools, especially within the teaching of **STE(A)M**—science, technology, engineering, arts, mathematics—subjects. Biomimicry can link biology more to STE(A)M education and also link science to students' everyday lives. The aim of STE(A)M education and also the education of biomimicry is to encourage students to take part in research, and start critical thinking and dialogues. Some benefits of inserting biomimicry into teaching: does not require new content to be added to the curricula; it provides a wide variety of academic ideas and subjects; academic ideas and subjects can be explored in connection with the engaging context of the natural world; it can be used in many subjects, so it can help to get nature closer to students. It can also help in improving creativity, problem-solving skills, self-reflection, and design-thinking skills.

Our modules can be used within education lessons and out-of-school activities. As we can learn from nature how to separate the world sustainably, it is also a great tool for educators for sustainability.

Within the **BioLearn** project, we developed 22 "universal" **modules** written in English and made for learning all over the world. On the national pages, there are some modifications according to national needs: the English page is made especially for the United Kingdom's schools; the other pages may contain fewer or more modules than the universal version. This article introduces the universal page of the project (https://biolearn.eu/universal-manual/), which is indicated with the logo of the European Union.

The modules have a unified structure. After the title, we find a subtitle with a challenge or question about what we can learn from nature.

The first part is an overview of the module:

- Summary:1–2 sentences about the content
- Biomimicry Principles: Relevant principles from the possible nine
- Learning Objectives: Direct goals and attitudes should be reached by the end of the module
- Learning Outcomes: What will the students achieve by the end of the module?
- BioLearn Competences: Relevant competencies chosen from a series of biomimicry competencies
- Summary of the Activities: Table with the main features of the activities (name, short description, method, duration, and location)
- Age Range: Recommended age
- Duration (Preparation, Activity): In minutes
- Subjects: Relevant STE(A)M subjects
- Keywords: Most important words of the content of the module

The second part contains the details:

- Background for Teachers: Some deeper background on the module and how to implement it.
- Activity Details
    - o Description: The activity itself
    - o Extension: Versions, other ideas
    - o Literature, Additional Information: Literature connected to the topic

  - o Location: Outdoor/Indoor
  - o Tools and Materials: Physical tools and materials needed for the activities; sources for research during the activity
  - o Preparation: Classroom arrangements, things to do before the activity
  - o Resources: Websites/annexes
- Teacher's Pages: Important materials for the teacher to show or read to students during an activity, solutions for exercises, and additional information.
- Student Worksheets: Easy to print/copy and give to the students.

The series of modules are divided into two groups: the "Introduction Modules" and the "Other Modules".""

Within the **Introduction Modules,** we can find the Biomimicry Principles, Marvellous Models, Introducing Biomimicry, and the Big Biomimicry Challenge.

Biomimicry Principles:

This series consists of 10 separate modules (an overview and the nine principles).

In the overview modules, you can find a presentation of the nine principles and a detailed description of the presentation. It is very important to see these nine principles clearly, as they serve as a sustainability guide to nature. At the end of the modules, students can discover the interconnections among the principles with the help of an activity.

The other nine modules are about the principles. In each module, you can find the slide and the description of the given principle from the overview for having iteration, one or two activities (mostly games) and a review for closing.

A brief summary of the games and activities in these modules:

1. Nature runs on sunlight

   Playing photosynthesis: A role-play about the process. The students will play the following roles: Sun, Water, Norman (the

Plant), Sunlight, Oxygen, Carbon Dioxide, Sugar, and Narrator. The play helps them to understand the process.

Modeling the reaction of photosynthesis: This is for students who have already learned chemistry. For preparation, teachers should make cardboard discs with the chemical symbols of the elements of photosynthesis (C, O, H) in the number of the chemical reactions and stick them onto kebab sticks. With the help of these sticks, students can model how photosynthesis happens.

2. Nature uses only the energy it needs

   Searching for energy forms: Students search for different forms of energy in nature with the help of small cards, which include words connecting to energy (e.g., insulate, cooldown, collect sunlight).

3. Nature fits form to function

   Recognizing forms and functions: Students observe different (natural and manmade) objects and detect what their function is while they use blindfolders.

4. Nature recycles everything

   Soil ladder: Students observe how leaves become soil after falling from the trees.

   Compost game: Students figure out what elements of waste can or cannot go into the compost.

5. Nature rewards cooperation

   Playing an oak forest: Students become elements of an oak forest (including non-living elements such as soils and sunshine) and see the interconnections among them.

   Honeybee dancing: Students mimic the dance of a honeybee and try to understand the meaning of the "dance steps."

6. Nature banks on diversity

   Playing a black locust forest: This is a very similar game to "playing oak forest in 5th principle, but the students become elements of a black locust forest (or any non-native forest with only a few species). As there are not too many species, more students get the same role cards, and they can observe that an ecosystem with fewer species can be more vulnerable.

7. Nature demands local expertise

   Playing beaks: The beaks of birds are adapted to the type of food they eat. Students try to understand this by using different tweezers for picking up different objects.

   Triangles: A simple way of arranging into a kind of order by following simple rules (and not a central command).

8. Nature seeks balance

   Playing deer and natural resources: A simple game to show how the deer population can reach a balance in different circumstances.

9. Nature taps the power of limits

   Harvest game: Students become members of different fishing companies and try to make a living from a lake with determined numbers of fish.

Introducing Biomimicry

A summary module for beginners. It shows some biomimicry examples, contains a card game with matching pairs, goes through the nine principles, and even shows the biology-to-challenge way of the biomimicry design process using the example of a seed.

Marvellous Models

This module focuses on the basic skills that are needed to be able to learn from nature. It deals with the first type of biomimicry design process—what can we learn from an organism? Students observe a chosen organism, write characteristics and functions about it, and try to find out what we could use from these features.

Big Biomimicry Challenge

This module shows the second type of biomimicry design process: from challenge to biology. Students can choose challenges connected to the Sustainable Development Goals (SDGs) and try to solve them with the help of nature.

Other modules

The other modules are mostly types of challenge-to-biology design processes: we have a certain problem and try to find nature-inspired solutions.

Packaging

This module is a very good and simple example of the challenge-to-biology design process. Like our food and many other products, every organism is packaged. Our skin, the armor of a crab, the peel of a banana, the shell of an oyster, the bark of a coconut, a pineapple (seed pack) and every cell in our body has its own packaging. We can discover how nature's different ways of packaging help us design solutions to our own packaging challenges.

Water Water Everywhere

A special type of packaging: the plastic water bottle. Students will explore how nature has a range of different abilities we can learn from. Students use the abilities of nature to inspire ways to address the challenge of plastic bottle pollution. By the end of the module, students will have created their own solutions to the challenge of plastic bottle pollution.

The Natural Economy

In the natural world, all materials are made using resources found locally; organisms do this at room temperature and pressure. Once materials are done with, they biodegrade back to their base element. This module explores how we can learn from nature to produce the materials we need sustainably. You can also find here how to make packaging from mushroom mycelium and the basics of the circular economy.

Buildings

Shelter, warmth and protection are all functions of buildings. In this module, students search for similar functions in nature and investigate how to use this knowledge to create better and more sustainable buildings: they plan together a building with features learned from nature. The module contains some ideas about how nature shapes strength and form, maintains constant temperature and cleanliness, removes pollutants, minimizes water use, manages natural light and ensures well-being.

Water Management in a City Park

Our challenges regarding water management are getting bigger and bigger. There are areas that suffer from extreme drought, while in other places, there are floods. Access to clean water is a major challenge for many people. In this module, students will discover these challenges and come up with solutions themselves based on how nature deals with water management. These challenges all come together in one case study: the design of a city park.

Healthy by nature

How do you become strong as a bear? Or how do you stay healthy as a fish? Do animals use medicine? What does nature do to stay healthy or to become healthy again when sick? What can nature teach us about health and the prevention of illnesses? We can learn these from this module with the help of games.

Plant Protection Inspired by Nature

In order to stay alive and healthy, we need to protect the environment, our body and our food (and crops) against damage. How can we do this without harming other living organisms on the planet? In this module, students will explore plant protection inspired by nature.

Adaptation to Climate Change

This module presents students with two challenges arising from climate change: one exploring the effects of warming in towns, and the second the flooding. Each challenge is designed for students to understand how nature deals with natural events and how we can learn from this to address climate change. Each challenge can be delivered as a discrete lesson or combined.

Cooperative Problem Solving

In these modules, groups of students train their observation skills, analogical and critical thinking, discussing and problem-oriented decision-making while learning from nature.

At the end of the modules, you can find the **Teacher Training Guide**. It is made for trainers, but as it contains useful background information about biomimicry and its usage in education, it can be used by anyone who is interested. The main chapters are About Biomimicry, Biomimicry in Education, Practical Approaches, Key Skills to Biomimicry, Introducing BioLearn Modules, and Setting up a Teacher Training.

Each European partner organized a **teacher training** based on the guide and the BioLearn modules. The aim of these trainings was to get teachers more familiar with the world of biomimicry and the modules and also to encourage them to use their new knowledge during their teaching practice. In addition to teachers, environmental educators and educators from zoos and the botanical gardens attended the training.

Within the project, we organized a **study trip** to the Netherlands for 55 teachers from the Czech Republic, Hungary, Slovakia and the United Kingdom. During the study trip, the teachers had the opportunity to meet with biomimicry experts, teachers who use biomimicry in their daily teaching, an education center with biomimicry toolboxes, ideas and exhibition and a carpet factory using biomimicry for working in a sustainable way.

At the end of the project, the European project partners organized **national conferences** to disseminate the results of the project.

However, the original BioLearn project was finished in October 2021; in November 2021, we were able to start a **new project** with the same partnership, named "**BioLearning for the Future - Smart by Nature**". The target groups in this project are the 9–12 year-old children and their teachers.

Within this project, we maintain the same website (https://biolearn.eu/) but will extend it with the content of the new project and a page for schools taking part in the program of the planned Biomimicry Days.

We developed 20 so-called biomimicry stories or exemplars—20 organisms we can learn from. These stories contain an introduction, a short activity or experiment, and a simple and more detailed explanation of how the world can be more sustainable by using the idea of the organism.

We also elaborate on the Biomimicry Learning Resources connected to the 20 biomimicry stories. It means that in each story, users will find a couple of activities that help in understanding the example. We also compile toolboxes to help in the delivery of the activities and a card game printed in the project countries.

In the second half of the project, we tried out the exemplars and the activities and organized the Biomimicry Days. These will be announced to schools, providing them with a series of materials and activities.

At the end of the project, we plan to have national workshops and an international online conference.

Findings/Experience and Evaluation:

Biomimicry and especially its usage in education are very new in the Central European countries of the project. These partners learned from the more experienced organizations from the Netherlands, United Kingdom, and the U.S.A. and the team elaborated on the educational modules and teacher training material together. The students of the pilot schools and the participants of the study trip, teacher training, and other dissemination events enjoyed this new method. We received very good feedback on all the programs organized within the project (work with pilot schools, study trips, teacher training, and national conferences). The more experienced partners also learned from the Central European environmental educators about outdoor and other methods that can also be used in biomimicry.

Some quotes from the participants of the programs:

Students:

"I will pay attention in nature what has nature already solved." (Hungary)

"When I am trying to think of a way of doing something, previously I would not have thought of using inspiration from animals, but now I would." (United Kingdom)

"I was amazed at how nature behaves, protects and works, and that even people in everyday life are inspired by nature."

"…nature has everything made for a reason." (Czech Republic)

Teachers:

"I latched on to the BioLearn project by letting the staff of Magosfa deliver presentations to students; then, in March of 2020, I took part in the study trip to the Netherlands. We could see many years of usage of biomimicry in schools, and we could meet Sam Stier from the USA. He is an excellent expert in biomimicry and can raise a problem from engineers to solutions of nature in a fantastic way. … I also attended the teacher training, and then I tried out modules with my students, e.g., Principle 9—Nature taps the power of limits—Harvest game. … In the next school year, I will use biomimicry for 11th-graders in science. BioLearn is a great tool in education for sustainability, as it uses practical solutions that work. Biomimicry complements the technique and technology-based thinking of our times."

"We received instantly usable teaching material, which builds upon the interest of the students and can be well fitted into school curricula."

"We were guided to a very exciting, interesting field with the help of the STEAM method, during which students can practice the methods of cognition of natural sciences on the basis of the phenomena and observations perceived in nature. The method, which involves science, technology, engineering, the arts and mathematics, supports innovation and a complex vision. (Hungary)

"It gave me ideas on interesting topics for teaching and practical exercises. It will definitely bring a lot of fun for the children."

"I really liked the seminar, especially for the provided clarity. It gave me many inspirations that can be used in everyday practice. Thank you!"

"Great, new, and inspiring to me. I will definitely use it in teaching biology and creating projects with students."

"They contributed new ideas—the STEM strategy in this activity combines the theoretical knowledge that students have acquired in lessons with a practical task." (Slovakia)

"I really liked the module's marvelous models. This really shows pupils how biomimicry thinking works. I want to use it in my biology lesson. Also, the online version was very attractive and good fun."

"I like the idea of biomimicry; it is quite new for me and the same for the students, but I would like to try several from the introduced modules." (Czech Republic)

"It helped me to think about what I teach and how to reconnect it back to nature. It was fascinating to me, so I'm sure it would be highly inspirational for students."

"I will use some of the ideas in my adaptations lesson with year 7, and I love the ideas for STEM clubs and careers."

"Absolutely fascinating, and it has inspired me to use it in teaching." (United Kingdom)

We can also tell that BioLearn was not an ordinary project—it really did not close by submitting the final report. It lives further, not only by implementing a new project but by many requests from teachers, schools and other educational institutes. However, biomimicry has really become a science, and few organizations deal with biomimicry in education. Requests to translate the modules into Turkish, French and Italian languages arrived, and other Central European countries indicated their interest.

Conclusion, Recommendations for Theory and Practice & References:

Biomimicry is a very good and inspiring method in education for sustainability. It widens the pedagogical approach of environmental educators and helps students find creative solutions to making our world more sustainable.

Within the BioLearn project, we elaborated:

- A completely new suite of learning resources (22 modules) for 11–16 year-old students linked with STE(A)M curricula to support biomimicry and nature-inspired learning, materials which have been piloted with pupils and demonstrated how they work in (and outside of) the classroom.

- A new website (https://biolearn.eu/) dedicated to providing teachers with the resources to explore biomimicry and nature-inspired learning.
- Teacher training workshops and programs to enable teachers to build confidence and competence in biomimicry and nature-inspired learning.

We encourage teachers and environmental educators to explore the world of biomimicry and use the elaborated teaching materials.

References

[1] An Idea That Stuck: How George de Mestral Invented the VELCRO® Brand Fastener (2016) The Velcro Company. https://www.velcro.com/news-and-blog/2016/11/an-idea-that-stuck-how-george-de-mestral-invented-the-velcro-fastener/

[2] Benyus, J. M. (2002) Biomimicry—Innovation Inspired by Nature. William Morrow, New York, U.S.A.

Sources of figures

[1] https://thinkjarcollective.com/wp-content/uploads/2014/12/Mestral-velcro-patent.png

[2] https://www.ubuy.hu/en/product/1V69YSO-velcro-brand-tape-5cmx1m

[3] https://www.dailymail.co.uk/news/article-4609128/Amateur-photographer-captures-kingfisher.html

[4] https://www.hogyan-kell.com/articles/japan-a-shinkansen-szuper-nagysebessegu-vonatok.html

[5] https://www.weforum.org/agenda/2021/10/how-mushrooms-and-microorganisms-could-transform-food-packaging/

[6] https://sustainable-nano.com/2018/11/28/nano-textiles/

[7] https://www.designnews.com/materials-assembly/water-repellent-coating-developed-less-harmful-alternative-natural-fabrics

[8] https://biomimicry.net/the-buzz/resources/designlens-biomimicry-thinking/

[9] https://biomimicry.net/the-buzz/resources/designlens-biomimicry-thinking/

Relevant congress subthemes

01 Action competence and key competences for sustainability
02 Arts, ethics, and EE
03 Business Sector and EE
04 Climate Change Education
06 Education for Environmental Citizenship
07 EE in Non-Formal Settings
08 EE in the Anthropocene
09 Outdoor EE: Bridging Nature Experiences with School and Home
12 Transformative, Transgressive learning in communities
13 Whole School Approaches to Environmental and Sustainability Education Cross-cutting theme Water for life

Chapter 6

Civic participation in rural areas during the pandemic

Reflections on the Community Castles Project

Elena Pagliarino[1], Bianca La Placa[2], and Gabriele Ardizio[3]

Abstract

The present study investigates a community civic initiative (the Community Castles Project) developed in the rural northwest of Italy during the COVID-19 pandemic. The project was built around the concept of physical places (two medieval castles) that can become community centers for cultural, recreational and decision-making activities.

This paper focuses on the narrative of the project, and its main objective is to identify which actors and processes are relevant to fueling civic participation in rural areas.

The data used to discuss key learnings were collected through in-depth interviews, focus groups, examination of project documents and observation of meetings.

The analysis of the data revealed that civic initiatives can be activated by external facilitators, but the early and active participation of community leaders is needed to achieve civic involvement and support the project. Local leaders can network with each other by means of a community board, which becomes a tool within their transformative and generative capacity to inspire and involve others in collective initiatives.

This article contributes to the literature on civic participation in rural communities by presenting a perspective new to rural develop-

1 Italian National Research Council, Research Institute on Sustainable Economic Growth, Turin, Italy
2 World Environmental Education Congress (WEEC) Network, Turin, Italy
3 Consortium Castel of Buronzo, Buronzo (VC), Italy

ment practices due to its development during the pandemic.

Keywords: rural areas, civic participation, facilitation, local leaders, COVID-19.

Introduction

Rural areas are undergoing profound transformations. On the one hand, their role in the conservation of natural resources and biodiversity and in the mitigation of climate change makes them strategic in the ecological transition (Wieliczko et al., 2021). On the other hand, a series of fragilities—such as low population density, population aging, depopulation, geographical marginality, scarcity of services and cultural opportunities—puts rural territories at risk of decline (Price et al., 2017).

Civic participation is considered crucial for the sustainable development of rural areas since the sustainability transition requires empowering local communities to choose the transition pathway most aligned with their current situation and development capacity (Chartier et al., 2021). Ostrom (1990) argued that a community, i.e., a social system capable of its own autonomous regulation, can collectively use environmental resources without necessarily depleting them, contradicting Hardin's "tragedy of the commons" (1968). Such communities build their identity on the natural capital of their territories and, in turn, add a cultural heritage to the physical places in a process that is expressed and synthesizes in the landscape (Agnoletti, 2014).

Civic participation is also said to be closely linked to a socially healthy, engaged, and integrated population (Putnam, 1993a, 1993b, 2000), and a high level of civic participation seems to improve the well-being of residents (Noguchi-Shinohara et al., 2020).

Nevertheless, the level of civic participation of Europe's rural inhabitants is rather low. Citizenship mainly takes the pattern of participation to elect representatives, while other forms of civic participation, including voluntary associations or political engagement, are much less common. There is a high degree of differentiation in civic participation across the European countries, with Italy belonging to the subset of countries with a medium level of participation (Starosta, 2010).

The presence of community meeting places (also inter-generational) can facilitate civic participation, i.e., the process of comparison, decision-making and planning of community initiatives. Community meeting places used to be a common feature in rural areas, but they have disappeared with depopulation and the progressive closure of shops and social and recreational services. In Italy, some policies targeting rural areas have tried to rebuild or, better, recover places that fulfill this function. Law no. 158 of 2017, for example, aims to create multifunctional centers where emporiums, bars, post offices, tourist information centers, free Wi-Fi areas, doctors and pharmacies are concentrated. Some bank foundations have also moved in this direction through specific tenders (e.g., AttivAree of Fondazione Cariplo and CivICa of Fondazione Sanpaolo) to encourage civic innovation by promoting cultural activities in physical places of aggregation already present in the territories and acting as poles of attraction. They are preferably places with a strong identity, which can be easily recognized not only by the local community but also by tourists, places with a historical and cultural value, managed by the community itself and taking on the function of 'community houses'. Scholars, practitioners, policymakers, and funding agencies are interested in how to foster civic participation in rural areas. Hence, understanding its dynamics, drivers, barriers, and good practices may help: i) scholars to better understand the theoretical context, ii) practitioners to implement more effective community-based interventions, and iii) policymakers and funding agencies to design and support targeted policies, programs and actions for the inclusive and participatory transformation of rural areas.

This paper contributes to discussions around this issue by sharing the experience of *Castelli di Comunità*[4] (CDC), a civic participation initiative to involve the communities living in two rural villages in Piedmont, North Italy. Several activities were planned to strengthen the local identity and sense of belonging to the community and to create the capacity to collectively design further civic initiatives. CDC is a pilot, exploratory project capable of contributing to investigating civic innovation in rural areas. Furthermore, its development is particularly interesting, as it was conducted between October 2020 and December 2021 in the midst of the COVID-19 pandemic, with strict prevention and containment measures in place, prohibiting or limiting the meeting and aggregation of people. In the next sections, we will

[4] *Castelli di Comunità* translates from Italian as Community Castles.

provide an overview of the theoretical framework and describe the community project. This is followed by a discussion of the method used to analyze the project. The findings are then presented, followed by a discussion and conclusion.

Theoretical framework

The project included an internal evaluation commissioned by the partnership and conducted by the authors of this article. The evaluation explored what worked and what did not in terms of resources and limitations of the partnership, critical issues encountered during the project, and opportunities for improvement. The analysis involved the people who played an important role in the project, offering a summary of their sometimes conflicting points of view that highlights the participants' understanding of the processes as well as their awareness of everyone's responsibility (Portales, 2019). Our purpose, therefore, was not to pass judgment but to contribute to the learning process of the partnership (Henry and Mark, 2003; Schön, 2017) so as to build a transformative capacity that supports the diffusion of a culture of improvement and the empowerment of rural communities (Díaz-Puente et al., 2007). A mix of approaches was used in the analysis. In detail, the Theory-Based Evaluation (TBE) approach (Weiss, 1997) was adopted in addition to the constructivist approach (Cronbach, 1980). According to the constructivist approach, attention is drawn to the transformation of a program during its development, arising from both the context and the interpretation of the actors involved in its implementation. The TBE approach focuses on the causal mechanisms linking the program and its effects (Stame, 2004), looking for the theory (or theories) underlying its functioning—"the series of small steps that lead from inputs to program outputs" (Weiss, 2007, p. 355)—through the definition of a set of assumptions on the relationships between strategies and results (Rossi et al., 2004; Lumino, 2013). Both approaches require stakeholder participation in the evaluation process and attention to the context, which comprises constraints and resources that contribute to the pursuit of results (Pawson and Tilley, 1997). The relationships between actions taken and results emerge through a variety of methods—including case studies, in-depth interviews with key informants, Focus Groups (FG), observation during meetings, document analysis and recording of events and activities—

and are discussed on the basis of the stakeholders' experiences and the literature (Butterfoss, 2006). The evaluation of a project fulfills a dual purpose: on the one hand, it allows improving it from a constructive perspective (formative); on the other hand, it makes it possible to review the results as a whole (summative), with a view to redirecting the program or, more generally, to achieving generalization, transferability and scaling across contexts (Scriven, 1996).

Before proceeding with the empirical part of the article, it is useful to explain what is meant by civic participation.

According to Putnam (1993a, 1993b, 2000), social capital, civic engagement, social inclusion, and civic participation are key to a healthy, engaged, and equal society. They are all seen as desirable goals and, consequently, encouraged in rural development programs (Anderson & Bell, 2003). They are also credited with facilitating rural development (Ray, 2000, Ryan et al., 2005; Curtis et al., 2014).

Social capital refers to dense networks of civic engagement that produce a capacity for trust, reciprocity and cooperation (Shortall, 2008), and it is usually only discussed in terms of its ability to lead to economic growth and a healthy democracy (Lowndes d Wilson, 2001). Civic engagement, instead, is not motivated by the goal of making a profi;, it can be individual or collectivee social or/and political, goal-orienteed or an end in itself (Fiorina, 1999). Civic engagement is the network of ties and groups through which people connect to one another and get drawn into community and/or political affairs (Ibidem). Both social capital and civic engagement need the participation of individuals or groups. Fiorina (Ibidem) argues that people participate when motivated to do so, and the reasons for non-participation may be the result of an active decision not to participate. Interest in participating and the ability to do so, together with the presence of social structures that allow for participation, determine the act of participation and, consequently, the social inclusion or exclusion of specific individuals or groups (Shortall, 2008).

In this article, civic participation is defined as a process where individuals, groups, and organizations choose to meet and take an active role in making decisions that affect them (Wilcox, 2003).

The civic participation initiative investigated: the CDC project

CDC is a participatory 'design and build' project piloted during 2020-

2021 in two rural communities (Buronzo and Rovasenda) in north-western Italy (see the map in Figure 1).

Figure 1: Geographical location of the case study. Source: Google Maps, 2023.

A bank foundation financed the project as part of a tender to encourage civic participation based on priority interventions in the cultural sphere and the enhancement of physical participation plaion. It intended to pursue this objective through a series of community aggregation, facilitation and training actions. The backbone of the project envisaged the participation and support of different actors in the local community, such as public institutions, associations, entrepreneurs, schools, young people, children and their families, and the elderly. At the design stage, brainstorming sessions and site visits were conducted to identify which stakeholders were more likely to be involved and to update the project plan so as to accommodate any changes needed. In this phase, it was possible to evaluate the inputs coming from different stakeholders, their interests and how they could contribute to the project's developmelf. At the implementation stage, the participatory approach was actualized in a series of workshops that took place throughout the project and saw the participation of the community board, i.e., a group of people representing the community.

The engagement of the community representatives made it possible to reveal the civic needs of the territory and its peculiarities (both its resources and critical issues), which the objectives, activities, and targets of the project had to address.

A visual representation of the project design idea is provided in Figure 2.

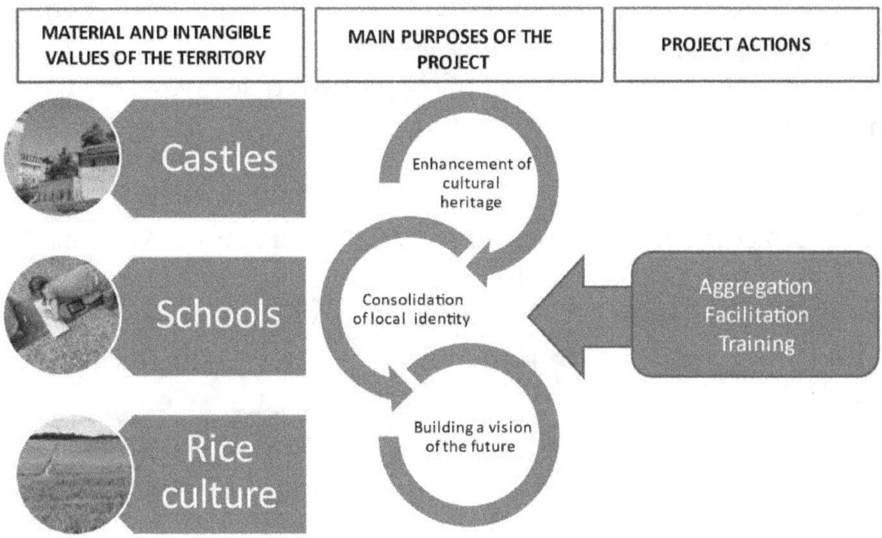

Figure 2: Overview of the civic participation process of the CDC project.

The geographical area in which the initiative was developed is in the heart of the so-called Baraggia, i.e., an environment of great natural value (moors and wetlands with outstanding plant and animal biodiversity, especially aquatic birds). Rice cultivation has been practiced for centuries and has defined its landscape and culture, so much so that there is talk of a 'civilization of rice'. From the period after World War II onward, with the reclamation of the Baraggia and the Green Revolution, rice cultivation became more and more intensive, requiring the heavy use of water and pesticides. Today, the conventional rice cultivation moion is starting to be questioned, thanks of a group of innovative rice farmers experimenting with alternative techniques, such as organic, biodynamic, integrated, precision, and green mulching.

At a more general level, the entire community, which has long been relying both economically and culturally on rice cultivation, is reconsidering its futre, with a view to ecological transition. Nonetheless, the available opportunities for civic participation are few and limited to recreational events (village festivals).

Therefore, the project intended to foster the civic participation of the community around an initiative to enhance the local cultural heritage (material and immateril), in order to develop a new rural culture.

The area's rural context and the challenges connected to it emerged during the participatory planning phase (Table 1) and were taken into consideration in the definition of the project objectives, activities and methodologies.

Table 1 - Summary of the characteristics and civic needs of the territory.

BURONZO AND ROVASENDA MUNICIPALITIES - BARAGGIA VERCELLESE		
CHARACTERISTICS		CIVIC NEDDS
CRITICAL ISSUES	RESOURCES	
C1 Small communities C2 Isolation C3 Neglect of the territory C4 Aging of the population C5 Agricultural economy concentrated in the rice sector C6 Environmental and social impact of intensive rice cultivation C7 Commuting C8 Youth unemployment C9 Lack of services, especially transport and Internet connection C10 Scarcity of cultural offers C11 Exclusion of young people from the life of the community	Legacy of the rice civilization: rice landscape and culture Sustainable rice farmers Castles Schools	CN1: Need to imagine the future of the community with a view to ecological transition CN2: Need to consolidate/redefine the community's collective identity CN3: Need to create places of cultural participation, especially for young people CN4: Need to create connections among territories

In an interview, the project coordinator explained: "One of the first civic needs is to overcome our rather strong cultural isolation. Due to the lack of local cultural opportunities, the main urban centers, which function as cultural attractions, satisfy this need. Thus, there is a sort of cultural commuting, especially by young people. In response to this need, the project aimed to provide opportunities for cultural aggregation at the local level and to create awareness in the community of the value of the local heritage, allowing sharing it outward. Therefore, the main objective of the project was raising awareness of the value of our cultural heritage, first of all, our two castles, then rice cultivation, above all sustainable, which contributes to creating the landscape of the Baraggia, and finally the presence of a school, or rather schools. These are the precious assets that had to be endorsed, first and foremost, in the eyes of the local communities so that they would gain an awareness of their presence."

The two tables below show the correspondence between civic needs and critical issues on the one hand and community objectives and recipients on the other.

Table 2 - Civic needs, general project objectives and community targets

Civic needs	General objectives	Targets
CN 1 & 2	GO 1 Fostering a new rural culture based on sustainability	Institutions, associations, businesses, schools, families, young people, tourists and tourism intermediaries (journalists, tourist and nature guides)
CN 1-4	GO 2 Fostering the hub of rice civilization	Institutions, associations, businesses, schools, families, young people, tourists
CN 1-4	GO 3 Understanding and improving current processes	Partnership and main actors involved

Table 3 – Critical issues, specific project objectives and community targets

Critical issues	Specific objectives	Targets
C2 & 10	GO 1 and 2 Promoting knowledge of the local area and cultural heritage	Schools, families, tourists and tourism intermediaries (journalists, tourists and nature guides)
C2 & 10	GO 1 and 2 Activation of civic participation	Institutions, associations, businesses, schools, families
C2, 10 & 11	GO 1 and 2 Activation of youth participation	Young people
C5, 6 & 10	GO 1 Disseminating skills for a new rural culture	Businesses, teachers, tourism intermediaries (journalists, tourist and nature guides)
C5, 6 & 10	GO 1 Disseminating expertise for organic agriculture	Businesses, especially rice farmers
C2, 5, 6 & 10	GO 2 Stimulating a community board of the rice civilization hub	Institutions, associations, businesses
All	GO 3 Monitoring and evaluation	Partnership and main actors involved

Methodology

This paper draws on reflection originating in fieldwork. The project was thoroughly investigated using a case-study research approach relying on qualitative data. **Midgley (2011) says that supporting evidence is often based on single case studies of interventions, and Meyer (2000) points out the need to consider what is unique in each intervention.**

The analysis was performed by combining:

- Document examination: all the documents relating to the tender through which the project was financed (Fondazione Compagnia di Sanpaolo, 2018 and 2021, Niessen, 2018) and the project documents were examined, along with the contents of the project website, posts on social networks (Facebook and Instagram), press releases in the local media, videos of the two community theater performances, and material produced by the spectators during the shows.
- In-depth interviews: Sixteen interviews were conducted with members of the partnership, the organization that funded the project, and the local leaders (Table 4). We adopted a discursive approach (Rapley, 2007) and facilitated the discussion by using an interview guide (Table 5), which consists of a sequence of topics to be covered and a list of open questions to be followed (Kvale, 2007).
- FG with the main actors of the project, which took place online on 10 Feb 2022, to check and discuss the results of the analysis (Bryman, 2008), following the interpretation of Corrao (2005), Zammuner (2003) and Albanesi (2005).
- Observation for the whole project duration by participating in nine meetings and four major events (kick-off meeting, family event, event for the two communities, and final conference).

Table 4. Overview of interviews conducted.

Members of the partnership	Funding Agency	Local leaders
1 Project Coordinator 1 Project Manager 2 Mayors of the two municipalities 2 Facilitators 2 Trainers	2 Donor Representatives	6 Local Leaders
8	2	6

Table 5. Interview guide.

Research Questions	Interview Questions
To what extent did the main actors involved understand the project (objectives, targets, activities and results)?	Based on your participation in the project, what were the objectives and targets of the project, what were the activities, and what were the results?
To what extent did the main actors involved participate in the project?	What was your role in the project? What activities/events/initiatives did you participate in? What were the reasons for your participation in the project?
What about appreciation, expectations, and outcomes?	What did you like, and what didn't you like? What did you expect from participating in the project? Did the actual participation in the project meet your expectations? If not, why? What was participating in the project useful for? Did participating in the project improve your knowledge of the local area and/or culture? Did you feel more attached to your territory/community? Have you improved your relations with other people?
Critical issues and suggestions for the future	What were the critical points of the project? What would you improve and how?

We used grounded theory to develop an understanding of the data collected. Grounded theory, in its latest evolution (e.g., Charmaz, 2006; Corbin and Strauss, 2015), is an interpretive method used to systematically analyze texts, such as interview transcripts and observation notes, in order to build theory concepts. This is done by reading the texts with specific questions in mind, extracting common patterns and themes and coding passages with keywords and quotes. Therefore, the interviews and other research activities were recorded, and the raw data were first paraphrased and then coded iteratively. Connections and relations were created among topics and clusters to give the data additional meaning and depth. In this way, a systematic perspective and lens were adopted. Lastly, the data were combined and clustered to identify key topics, which are presented in the Results section as a SWOT (Strengths-Weaknesses-Opportunities-Threats) matrix.

Results

Strengths

Local actors as precious allies

From the analysis of the project, it emerges that its most important strength was the involvement of local actors. Representatives of public institutions (the two Municipalities) and voluntary associations (Pro Loco[5], library, parish youth club, festivals and culture committee, youth group, etc.) and some members of the local group of sustainable rice farmers were involved.

The project coordinator described the experience as follows: "The role of some actors was decisive. I'm talking about some local catalysts, 'indigenous guides,' you might call them. This idea of the project proved successful, involving local people who, once they were trained and understood the objectives of the project, were able to involve others in their associations, acting as leavening agents in the area, so to speak. A clear example of this dynamic was the involvement of the Pro Loco manager, who, at a certain point, actively participated in all the meetings; she became very knowledgeable about the project's objectives and activities—and from then on, the Pro Loco began to respond. It can be said that she fully embraced the spirit of the project and played this very role of catalyst, of indigenous guide. So, she can be our local Pocahontas, a lady of a certain age who has proven very valuable."

According to Arzeni and Storti (2021), "the impoverishment of human capital represents a point of weakness of rural communities, but the demographic rarefaction, however, has the advantage of making it easier to identify and involve subjects in the area who are interested in its development to outline and share with them an idea of the future (multi-actor approach)." In small municipalities, it is easier to approach local leaders because "we all know each other and it is enough to make a few phone calls, but also because they are usually the ones who make things happen" (Interviewee 5). Indeed, these people are very much involved in the community, both in associations and the local government, with more than one civic office or responsibility.

[5] Pro Loco is a tourist promotion association.

The personal characteristics of these individuals (age, gender, family status, profession) are heterogeneous, but what seems to bring them together is their civic passion, active participation and commitment to their community.

Their involvement was essential for carrying out the activities and involving the communities, and it demonstrated that a participatory approach is effective for the success of a project.

The experience of the community board was formative for the participants and generated new initiatives. For example, during the board meetings, the capacity for collaboration of the rice farmers emerged beyond any specificities. According to them, living in the Baraggia territory represents a limit from an agronomic point of view but a marketing opportunity if appropriately exploited. The goal of promoting the area was keenly espoused by the producers. At the beginning of the project, the rice farmers had expressed the intention of setting up an association for the promotion of organic production and the protection of the Baraggia area, with its unique peculiarities. This happened in March 2021, with the establishment of the "Rice Organic District". The space for reflection allowed by the board and the skills training delivered through the webinars supported and facilitated this initiative.

Yet, the coordinator of the project was of a different opinion since, in his view, the role of entrepreneurs was marginal compared to expectations: "In my opinion, the project was based on a false premise, that is, on the presence of a cohesive group of rice farmers. In reality, these producers have very heterogeneous views on rice cultivation and find it hard to converge on common themes. The project intended to support them in communicating with and presenting themselves to the outside world. By telling others who they are, they would have the chance to understand who they are. But it became clear that they cannot tell others about themselves due to their lack of identity."

As for plans, the rice farmers are keen on expanding the group and collaborating with other businesses, not only in the rice sector, to build a community economy that includes citizens and consumers (following the guiding principle of proximity), to make them more aware of the producers' commitment to environmental protection. This project revolves around rethinking the civic uses of the castles. According to an entrepreneur, "The cultural purpose alone is not enough. It is necessary to think of castle-related products and services that have a market. This can only be done with the collaboration of private individuals." A municipal official also believes that the castles are a key point to work on in the future, creating a process of integration and dialogue with the representatives of the production sectors.

In conclusion, the presence of a number of local entrepreneurs who are innovative and active seems to be a resource for the area. The project tried to animate it and partially succeeded, but much work remains to be done, and in this regard, the municipal administrations have an important role to play in continuing down the path paved by the project.

Innovative activities, in person

Some new activities for the community were another strong point of the project. In particular, the theater performance experience was very successful. The communities felt involved both in the show and the preparatory work, which consisted of meeting the communities themselves and collecting anecdotes about local characters, events, etc. The interviews revealed that the theater activities were perceived not as simple entertainment but as a real opportunity for civic participation and to reflect on one's collective identity and future. This is how one of the local leaders recounted the experience: "The experience of the community theater was a success. At that moment, we felt like a community. Some unpleasant things were also said about us, but lightly and ironically. Initially, the actors wanted to go house to house to collect the stories, but we were against it, also given the COVID situation. We are a country of people who are always willing to welcome others into our homes, but no one can be trusted to open their door to strangers. So, we organized a meeting in the hall of the castle, and then we encouraged all the others to participate and tell anecdotes from their past lives. It was a wonderful experience; it is an idea to be repeated on other occasions." (Interviewee 6)

The urban sketching initiative was also appreciated. It included an observation and drawing workshop accompanied by a guided tour of the area. An illustrator taught the participants to observe and draw while a guide explained what was being observed.

"The illustrator taught us to observe. It was a way to see the things you see every day through new eyes. What you normally don't notice anymore. Furthermore, drawing is an expression of the self and enables you to make what you see and draw your own. Drawing allows you to delve into what you are drawing. This has allowed them to reclaim their territory." (Interviewee 2)

More generally, the interviewees showed greater appreciation of all face-to-face aggregation activities that gave them the chance to meet in person. After months of closure due to the pandemic, this modality also had a liberating meaning for the participants: "I liked the theater a lot; after two tough years due to COVID, it was a return to normality." (Interviewee 6)

Weaknesses

The pandemic

The pandemic heavily affected the smooth running of the project, which was strongly based on a participatory approach, attendance, the use of non-virtual physical spaces, and conviviality in meeting other citizens. Furthermore, many training and facilitation methodologies, from the teacher training course to the theatre show, were designed to be applied in person to enhance the relational dynamics among the participants. COVID-19 required a rethinking of the methods and timing of the activities so that those planned between the end of 2020 and the spring of 2021 were moved online.

The training courses were not too heavily affected by this change. The one for teachers was possible thanks to the fact that after a year of working from home because of the pandemic, they were accustomed to online tools. The course for journalists, tourist guides and nature guides benefited from being moved online, as this enabled wider participation of people from all over Italy. Instead, the cultural initiatives planned to promote knowledge of the area and facilitate citizen participation were only partially replaced by virtual events (visits to castles and exhibitions), but most were suspended until May 2021 and then concentrated between spring and summer 2021, when it was possible to resume face-to-face activities.

The fact that many activities were moved online highlighted one of the critical issues of rural areas, namely the weakness of the Internet connection, which made it difficult to take part in meetings and events. However, COVID-19 also made it possible to experiment with online relationships for previously unthinkable activities and targets. For example, some senior members of the community board said that they were happy to learn how to use digital technologies to participate in the meetings envisaged by the project. "Thanks to COVID, the initial lack of digital skills of some community board members was overcome. We began to do online events to promote the area, for example, those for journalists, using innovative techniques: live virtual tours were created, with a guide who visited and showed the participants around the castles while they watched online." (project manager)

During an interview, the coordinator of the project summarized what happened as follows: "On the one hand, the shift to online activities, for example, remote meetings, allowed people to discover the potential of online tools, also for the elderly, who would normally never need to resort to technology to meet. However, it did present some difficulties. The experience of remote meetings was less pleasant and less effective overall. It must be said that if the experience of participating in a project is pleasant, this creates further participation. The feeling of satisfaction in participating makes one want to repeat the experience. Instead, the constant schedule changes created confusion and frustration. We had to reconvert actions conceived in presence to a different mode and then revert back to the initial mode..."

And yet, while stressful for the partnership, this continuous adjustment of activities made it possible to test the strong capacity for resilience of the partnership and the community. In the lockdown period of autumn 2020, the project had just started and could have been suspended. Yet, the funding body decided not to put ongoing projects on hold. "At a time when everything was closed and all activities were at a standstill—explained an official from the funding body—the projects we had started were the only opportunity for civic participation for many communities. For this reason, we asked the partnerships not to suspend the activities whenever possible." This decision was particularly welcomed by the communities involved. One of the participants explained his experience of that period: "There was a very difficult psychological climate, and being able to do something anyway was very important; we didn't stop!" (Interviewee 7)

The exclusion of some community groups, young people, and schools

Although the involvement of local leaders was one of the project's strengths, the engagement process left out some components of the community that were regarded as priorities, namely young people and teachers from local schools.

As for young people, the activities intended to facilitate their civic inclusion (Future Lab and young board) failed. After the first meeting was organized to present the Future Lab, which also featured a buffet and DJ set but was attended by only a small number of young people, the other three scheduled meetings were canceled because the participants said that they would not attend.

According to the interviewees, the exclusion of young people from the life of the community and their commuting to urban centers to meet educational, recreational and cultural needs are long-standing critical issues. Starting from a certain age, around the end of middle school, young people 'disappear' from the community because there are very few opportunities to get together. Here is how some of the interviewees described the situation:

"It is very difficult to bring people together, even more to involve young people. They are few. I'm having a hard time getting five people to participate! There is nothing here that binds young people except the parish. Everyone goes their own way. That's why the youth activity didn't work." (Interviewee 2)

"The children's response was better. Engaging children together with their families is easier. Young people, on the other hand, can't wait to get out of the village." (Interviewee 1)

"Those between 18 and 40 are excluded, they don't respond, they are not part of the community." (Interviewee 3)

"In the youth group, it's always the same people. Many work, others go to university. Between the ages of 20 and 25, I don't know what interests them because they don't participate. We had a youth party for 3-4 years, but it was always the same people. It's a shame because the Future Lab would have been useful for the future." (Interviewee 4)

Young people are not used to civic participation and lack the experience and skills to exercise it. A facilitator commenting on the Future Lab experience confessed: "Perhaps it was the first time they had heard of an initiative of this type. I felt some sort of cultural extraneousness in working on the future. They are probably not ready for initiatives of this type." This comment confirms the importance of a tradition of civic participation (Putnam, 1993a, 1993b).

Some local leaders admitted responsibility for excluding young people: "The Future Lab was a flop because young people do not want to collaborate; they are not willing to commit. But it is also true that we have failed to attract their attention, to hook them. Perhaps it would have been good to involve some young parents with teenage children, or perhaps not, as they might already be worried about them because of COVID, but we can try again in a more targeted way." (Interviewee 6)

Another respondent explained: "We need to ask ourselves whether we have communicated with them correctly. A greater presence on-site would have been useful. Physical presence matters. The communication of the project was mainly entrusted to tools such as social networks, the website and press releases. In addition, posters, brochures, and postcards were distributed locally to advertise individual events. This type of communication was not effective enough. It would have been better to integrate the tools used with word-of-mouth and personal invitations. A broader and more timely communication action could probably have improved involvement. We should have sent an invitation to all the young people, to be more inclusive. The Municipality could have provided names; then a municipal messenger could have delivered the invitations. The people from the Municipality gave their availability. Instead, only posters and postcards were used..." (Interviewee 5)

As far as schools are concerned, during the planning stage, they were considered a priority area for intervention and collaboration owing to their strong presence in both communities. Teachers and children were regarded as the main targets for both the actions to promote knowledge of the local cultural heritage and those to facilitate civic participation. Nonetheless, in both cases, the participation of schools was limited. In particular, the planned collaboration with the Institute from which the schools of Buronzo and Rovasenda depend was terminated due to a lack of interest on the part of the new manager, who took over from the one with whom contacts had been made in the initial stages of the project (coordinator of the project and project manager). Although not encouraged by the management, the teachers of the two communities could have participated in the training course for teachers *Planning the future for aware citizenship*. On the upside, many non-local teachers enrolled (this was possible due to the course having moved online) and participated actively, creating excellent local development projects, but none of them were from the project area (interview with the facilitator of the course). According to some interviewees, the teachers were not interested in the online course because they were "tired of distance learning". Also, in recent years, except for a few rare cases, there has been a lack of teaching staff linked to the local area and willing to collaborate on "outside school and school hours" projects. "The idea of starting from the schools is right because we need to rebuild the mentality of the village, but among the teachers, there is no longer that spirit of curiosity about what the village is." (Interviewee)

Threats

Facilitation without empowerment

The exclusion of young people and schools can most likely be ascribed to the fact that the project was devised outside the target area despite the efforts made to encourage local participation. In many cases, rural areas do not have the resources and planning skills to access funding tenders. Indeed, the foundation that financed the project admitted that the project proposals received from small communities in rural areas were, on average, of lower quality than those received from urban communities (interview).

According to several authors (Arzeni and Storti, 2021; Bosworth et al., 2016; Bock, 2015), in rural contexts, the integration of skills from outside the local territory into the network of relationships becomes almost indispensable from the perspective of active participation aimed at co-planning the development path. The use of external facilitators is considered necessary and welcomed if it encourages local partnerships in the planning and implementing the actions. It does not contradict bottom-up design but rather facilitates it.

From the words of the manager in charge of the course for teachers and the Future Lab, an interpretation emerges as to why the project started from a facilitation position but missed the empowerment goal. "Local development projects must take shape and be very well rooted in the local territory. Instead, this project was presented to the territory and experienced as a foreign object, descended from above. A lot of engagement work was done, but not very effective. The facilitation didn't work. More work on the territory would have been needed to engage local actors. In fact, projects need a strong and proactive local component, people from the area who get involved because things have to be pushed. [...] Participatory processes can be done anywhere, but some basic liveliness, which you can lean on, is needed. In this case, there is little cultural life and few associations. [...] How do we get to do these projects? You see an opportunity, you have a deadline, hear from local interlocutors and create the project with them. Afterward, once the project is financed, territorial educational work is needed in advance in informal meeting places if there are no poles of attraction. One needs an opportunity to re-share goals. The pandemic took that away, and we just kept going forward. Perhaps the projects should have been extended..."

The project coordinator expressed similar ideas: "On the part of the local actors, there was availability and openness, there was also an active role in socio-cultural animation, but there were difficulties on the proactive side. They asked us: «What should we do? Do you have any idea how to start?»" The proposals were difficult to implement because of uncertainty about the future. In this project, the proposals had to come from us because a bottom-up approach was not feasible in our operating circumstances. There is extreme inertia; they are deprived of this point of view. It was a pioneering initiative. More inclusive and bottom-up dynamics are possible but at a later stage."

Legacies

The theme of the legacy left by the project was very important to the client and represented one of its main expectations. It was hoped that the project would make the local communities autonomous and capable of continuing the project initiatives after it ended and in the absence of financial support from the client. The partnership intended to 'pass the baton' to the local community, institutions and associations. For example, project partner municipalities were expected to include CDC objectives in their policies and implement in-house management of related initiatives. In this sense, the project left a perception of feasibility and appreciation of many initiatives. According to one of the two mayors, the project has given the target communities "greater awareness of our identity and our territory and a less parochial vision." The project continued the mayor, served to fuel "the perception of the beauty that surrounds us, this beautiful land, this unbelievable place we live in." It made it possible to successfully test a cultural collaboration between two Municipalities that, despite being neighbors, had never before carried out common initiatives. The other mayor commented on the experience: "Beyond political affiliations, we recognized each other and acknowledged our long-standing friendship [...] the project united our two Municipalities at a time, that of the pandemic, in which everyone was divided. We collaborated in the enhancement of the territory."

Other interviewees had the same perception. "The collaboration between the two municipalities worked; they are two similar, though not identical, realities; the two mayors are in harmony, we have things that unite us, and I think it will continue in the future." (Interviewee 3)

"The project idea of involving the two castles, the two communities, is undoubtedly good and should be extended to other neighboring Municipalities. These are small Municipalities, around 500 inhabitants, but there are also smaller ones, between 200 and 300 inhabitants, so there is an advantage in uniting, even though it is difficult to overcome a certain degree of parochialism. However, in a few decades, there will be a single Municipality or else a single community." (Interviewee 1)

The awareness that collaboration generates added value in relation to civic participation not only in the two Municipalities but also in the surrounding area took root during the course of the project, and all the actors involved expressed their intention to continue in that direction. This certainly represents an important result in terms of legacy, considering the strong parochialism that characterizes the two Municipalities.

As regards the continuity of the rice hub, it aimed to be a long-lasting space for reflection and planning for the future of the community, "a meeting point for all the players involved in the culture and citizenship of Baraggia [...], which will host events, meetings, workshops, a documentation center, institutional and civil society activities, to promote the local production fabric and the food and wine heritage" (from the CDC executive project). The project established the management board of the rice hub, facilitated its coming together with regularly scheduled meetings and trained its members through courses designed to develop their knowledge and skills. However, the continuity of this space for civic participation is not viewed with optimism.

Within the partnership, both the coordinator and the project manager said they were skeptical about the ability of the local communities to proceed autonomously in the planning of civic initiatives. According to them, providing motivation and external facilitation are still needed. Reflecting on the project's results, the project manager stated that it was probably too complex compared to the local expectations, which, in her opinion, are simpler. In the project, there was a sort of anticipation of needs, which is a suitable challenge for local development projects, but perhaps it went a little too far, proposing activities that were much too distant from the needs and capacities of the local communities. As for future planning, she hopes for simplification and the downsizing of the objectives and activities to make them more usable and concrete.

According to the coordinator, "The needs are there, but they are not perceived; it is a culturally depressed area. We must not forget that until the 1950s, there was no economic development and even after that, the economy was predominantly agricultural, so not much schooling was required. There is a lack of previous cultural experience." "The mayors—he continues—did not perceive the added value of the project. They did not give value to cultural animation. They told us things like: "We are putting in the co-financing money; what are you putting in in exchange?" These are people who expect concrete, practical things that can be touched. Sometimes, they would look at you as if you were speaking a foreign language. The things they most appreciated were those in traditional formats: guided tours of the castle, parties..."

As far as the two mayors are concerned, they understood the organizational commitment and the quality of the cultural events that were organized, but they did not seem to fully perceive the meaning that the activities should have had. Commenting on the outcome of one of the events, a mayor quipped: "We are well able to organize a party on our own." It should be remembered that, in the intentions of the project, cultural events were not an end but rather an instrument of civic participation, functional to the coming together of the community and the rediscovery of its cultural heritage and identity. Similarly, the educational and facilitation activities of the boards were to support the process of designing and building the future of the community (civic imagination). If this awareness was not fully expressed by the two mayors, who were also project partners, it seems difficult to envisage any project continuity after its end, at least with respect to the less "concrete" civic development activities. Other interviewees showed greater awareness of the intangible objectives of the project while admitting that further support from the outside would be necessary for the continuity of the processes of civic imagination.

Opportunities

The greatest opportunity for the project lies in its coherence with rural development policies. These include the well-established actions of the European LEADER program, the more recent ones envisaged by Italian Law no. 158 of 2017 (Measures for the support and enhancement of small municipalities, as well as provisions for the redevelopment and recovery of the historic centers of the same municipalities), known as the "save the villages" law, along with the initiatives for inland areas with major geographical or demographic disadvantages envisaged by the National Recovery and Resilience Plan (PNRR), within the European Next Generation EU Programme.

Since 1991, the LEADER program has promoted local development in rural areas with an endogenous, place-based, bottom-up and multi-sectoral approach, supporting local partnerships (Local Action Groups or LAGs) in gaining control of the development process (Barke & Newton, 1997). The LAG board is the decision-making authority of the partnership and includes members representing the different groups that make up the community. The empowerment of local actors and the protection of the rural environment are among the necessary conditions for developing rural areas, according to the SHERPA project, which contributed to the preparation of the long-term vision for rural areas of the European Union (Chartier et al., 2021).

All these elements are present in the CDC project.

National law no. 158 of 2017 aims to rebuild or recover meeting and gathering places for the community, facilitating civic participation. The National Strategy for Inland Areas, an intervention line of the PNRR, also finances actions to construct or regenerate social infrastructures (educational and socio-cultural spaces, social and healthcare facilities) and provide social and cultural services.

Finally, some Italian banking foundations have also moved in this direction through specific tenders (e.g., AttivAree of Fondazione Cariplo and CivICa of Fondazione Sanpaolo) that aim to encourage civic innovation by promoting cultural activities in physical places of aggregation already present in the territories and acting as poles of attraction.

A range of initiatives are converging on this objective, which was also championed by the CDC project.

The client of the CDC project tends to support projects that can grow, have future developments, and are a model that can be replicated in other areas. In this sense, CDC represents an exploratory, pilot, experimental project which, after learning from experience, can grow, if not completely autonomously, certainly with a more bottom-up approach. An official from the funding body explained: "There is a lot to do in rural areas in terms of participation and civic innovation. Rural areas lag behind urban centers in terms of planning. In fact, among the projects we funded, CDC was the only one that took place in a rural area."

Discussion and conclusions

The construction of a common vision for the future

The construction of a common vision as an essential methodology for sustainable development was already considered a priority in Local Agenda 21, launched by the United Nations Organization in Rio de Janeiro (1992). In Agenda 2030 (United Nations, 2015), two objectives—number 11, Sustainable cities and communities, and number 17, Partnerships for the objectives—provide for actions for the construction of a common vision and the necessary capacity building.

Perna (2002, p. 157) explains that "the process of identifying or even just imagining the common good is threatened by the corporate struggle and by the fragmentation of interests and demands in the face of a reduction in available public resources". "A common path—he continues (ibid., p. 158-160)—is not the sum of individual interests, often in contradiction with each other, but a project that starts from a common vision of the future which, in turn, rests on a strong sense of belonging to the territory, and why not of pride of belonging, on an affective bond of the inhabitants with their own land, on cultural autonomy, digging into the symbolic dimension of places." In this sense, the castles of the project are symbolic places around which to create opportunities for aggregation and build the process of enhancing the cultural heritage, strengthening identity and planning actions that anticipate and satisfy the local cultural needs. In the CDC project, they are perceived as real 'community houses' (Interviewee).

The process of building a strong and shared common vision passes through discussions with the different actors of the local communities and associations, including youth associations, and entrepreneurs... (for example, in Bobbio, 2004). This point was central to the project's approach, which sought to involve all social partners by using innovative tools and methodologies, such as the community board for adults, the Future Lab, and the young board for young people.

Arzeni and Storti (2021) argue that changes at the local level must be supported by investing in the involvement of local actors in training, not only young people in experimenting with innovative practices to support local processes, facilitating the creation of local networks. They also highlight that the use of these approaches is time-consuming. In this regard, the CDC project suffered from a short timeframe (18 months), further compressed by delays and schedule changes due to the pandemic.

According to De Rita and Bonomi (1998), it is necessary, together with the various actors of the community, to identify which ties with the territory are strongest and establish which local specificities, environmental, landscape and cultural assets one wants to defend at all costs, what should be off-limits, so that these assets are preserved for future generations. Around these, a shared strategy of self-sustainable local development should be built, unleashing imagination and creativity and fueling a collective dream. Thus, a local development project takes the form of a territorial project, a long-term scenario and a system for building strong links among local actors (Magnaghi, 2000).

The process just described is triggered from below, but according to several authors (Arzeni and Storti, 2021; Bosworth et al., 2016;

Bock, 2015), in rural areas, it can be facilitated from the outside, like in the case of the CDC project. This can be done by means of initiatives and events promoting the local territory and having high-added value (for example, training and education). Bringing in individuals from outside the territory (facilitation professionals, trainers, experts, researchers) feeds the process of reflection, comparison and mirroring and, at the same time, creates social bonds and friendships, providing concrete and pleasant opportunities for the sense of belonging to be expressed.

In the CDC project, the social partners involved were:

- public institutions (the two Municipalities);
- local schools (teachers, children, and families);
- social and cultural associations;
- the business world;
- young people.

In each area, with a view to reciprocity with the two local communities, the idea of the project was to offer support in the organization of training and promotion activities and events to help the local actors meet. Then, the process of dialogue, reflection on their vision for the future and planning should have occurred spontaneously, facilitated by the project.

The strong values to be preserved, very clear to all the partners and actors involved, were:

- The beauty and uniqueness of the landscape, the expression of the harmony between the action of man and nature, the result of the widespread cultivation of rice in the Baraggia environment at the meeting point between the mountains and the plains;
- the local history and culture, which, although similar to those of other rural areas, features micro-diversities due to the legacy left by the civilization of rice and the deep-seated parochialism typical of these territories, where castles replace bell towers;
- The strong presence of schools (from kindergarten to lower secondary) guarantees the existence of a population of children, teenagers, and families, in contrast to the depopulation process typical of most rural areas.

According to Za (2001), a local civic participation project is also pursued through the recovery of memory, not seen as nostalgia but as a technique for remembering to experience one's territory. This concept and the idea of leveraging creativity and imagination were considered, especially in the facilitation process using the theater methodology.

Rural areas and civic participation

The literature has highlighted the heterogeneity of conditions found in rural areas and, at the same time, the set of similarities that characterize them (Li et al., 2019). Rural territories are fundamental in pursuing the goal of ecological transition (Wieliczko et al. 2021). They play a crucial role both in preserving natural resources and biodiversity and in mitigating climate change. In this sense, the territory of Buronzo and Rovasenda has very high environmental value but also elements of fragility that risk compromising it. The uniqueness of the Baraggia territory has been recognized by establishing protected areas for the conservation of habitats for plant and animal species. Rice cultivation, practiced since the 1950s, has supported the conservation of typical wetland species thanks to its system of irrigation canals and the practice of submersion. At the same time, however, intensive rice cultivation has led to increasingly negative environmental consequences, including soil pollution and contamination of surface and deep waters, with risks to human health and loss of biodiversity (ISPRA, 2018). Submerged rice is also a source of methane emissions, one of the most harmful greenhouse gases. Current rice cultivation, therefore, requires rethinking by all the players in the rice-growing chain, but also by the communities living in the rice-growing area, still heavily dependent on the rice economy.

Civic participation is considered crucial for the sustainable development of rural areas, as the transition towards sustainability requires local communities to choose the transition path most in line with their current situation and development capacity (Chartier et al., 2021).

Farmers have an important function in this process. From a removed position, they have now taken on an active role in social change and innovation, contributing to territories' social and economic revitalization (van der Ploeg, 2008; Milone and Ventura, 2019). Various authors (Sanchez-Zamora et al., 2014; Salvatore and Chiodo, 2017) have highlighted that the consolidation of the intersectoriality between agriculture, tourism and conservation of natural resources might favor a condition of pre-resilience, providing local communities with the strength and ability to face crises and changes more successfully. In the project's target area, a group of innovative and sustainability-oriented farmers have started a process of dialogue and collaboration, both internally and with other businesses and local administrations, aimed at enhancing the territory with a view to cultural promotion and ecological transition. The project facilitated this process in ways (training activities and spaces for civic imagination) that can be replicated autonomously by the local community.

A general challenge for rural areas is that of demographic change due to urbanization, i.e., the migration of the younger population to urban centers, resulting from a decrease in rural population and its aging (Féret et al. 2020). According to Price et al. (2017), due to these phenomena, rural areas run the risk of marginalization in the political agenda, accompanied by reduced services, businesses and social capital, triggering a spiral of decline.

Civic participation is considered one of the goals of the long-term vision for rural areas in the European Union (Chartier et al. 2021). Noguchi-Shinohara et al. (2020) investigated the relationship between civic participation and the health status of people living in rural areas characterized by population erosion and aging, highlighting a positive correlation between civic participation at the local community level and good perception of one's health. Nevertheless, the level of civic participation of Europe's rural dwellers is quite low. Citizenship mainly takes the pattern of participation to elect representatives, while other forms of civic participation, such as voluntary associations or political engagement, are much less common. There is a high degree of differentiation in civic participation across the European countries, with Italy belonging to the sub-set of countries with a low to medium level of participation (Starosta, 2010).

High levels of civic participation are linked to a socially cohesive and integrated population, but it has also been observed that participation increases with socioeconomic status and that some social groups tend to be excluded (Kleiner, 2020; Roßteutscher, 2005; Warren, 2001). Kleiner (2021) points out that, in urban areas, social rooting occurs only for those who participate, while those who are not members of associations are also excluded from social networks and opportunities. In rural areas, however, the risk of exclusion due to the lack of social networks is cushioned by what Putnam calls "bridging social capital." Since there are few opportunities for civic participation, individuals from a household may be members of more than one association and/or different associations. Anyone not a member of an association is included on a social level not because they take part but because another member of their family, who is active in the association sector, connects them with other individuals in the community. Thus, in rural areas, voluntary associations help to increase the community's social cohesion.

The key role of local community leaders is also well-recognized in the literature (Li et al., 2016; Vercher, 2022). Capable, charismatic leaders and resourceful entrepreneurs are needed to inspire other community members, negotiate deals and implement projects (e.g., Horlings, 2015). A less investigated aspect, however, is how much the training of such civic actors benefits community development. Godwin et al. (2012) studied the results of some American educational programs to develop leadership skills in rural communities. They stressed that the programs were successful in shaping individual knowledge and skills useful for people to act as leaders in their communities and in developing the community in terms of social capital and improvements in five other community areas: cultural, political, infrastructural, natural and financial.

A separate matter is that of the civic participation of young people. Some studies have shown that young people are less engaged in the community than the older generations (Adler and Goggin, 2005) and that, just like among adults, youth participation increases as individual socioeconomic status increases (Pritzker and Metzger, 2011). In rural areas, youth civic participation is limited compared to urban centers, but rural territories would be the best environment to learn civic and political engagement (Lay, 2006; Elder and Conger, 2000). Social initiatives dedicated to young people in rural areas focus on care aspects, while there is a lack of projects aimed at facilitating their connection with the community and their civic engagement (Checkoway and Gutiérrez, 2006).

Although research on youth civic participation in rural areas is still scarce, some facilitating elements have been identified, like support from adults in the community and the presence of schools and other associations that involve young people in extracurricular and church activities (Pritzker and Metzger, 2011). At the same time, youth civic participation is hampered by the scarcity of civic opportunities, homogeneity, and distance from diversified opportunities, which prevents easy access to them (Ibidem). Furthermore, due to many young people moving away or commuting to work and study, the young adults who remain in rural areas tend to be less educated and less civically involved, causing the absence of positive civic role models for the adolescents in the community (Johnson, 2006). The civic participation of youngsters should be facilitated through projects dedicated to them, specifically and directly inviting their contribution. The projects should preferably be developed together with adult leaders and local organizations that have a facilitating function (schools, churches, recreational and sports clubs and associations), with the involvement of the young people themselves in the planning of proposals, so as to develop community governance skills in them (civic imagination, decision-making, leadership…) (Pritzker and Metzger, 2011).

In our project, civic participation was facilitated by local associations and, in particular, by some local leaders. The training and facilitation activities provided were aimed precisely at involving these local actors, who are among the most active at the civic level in the main areas of the community (institutional, association, educational, youth and entrepreneurial) so that they could become catalysts for the civic planning and cultural development of their community. The validity of this idea finds confirmation in the literature, as we have seen. In its application, the engagement work was not sufficient to identify such leaders for all social partners, leaving out young people and teachers. In particular, as for the young people, the failure of the initiative is ascribable to the critical issues typical of rural contexts, which are also confirmed by the literature on the subject.

The analysis of the project experience suggests that it should continue with a greater focus on developing leadership and planning skills among the most important civic actors of the community, including young people, to whom specific initiatives must be dedicated.

According to the results of the SHERPA project on the vision for rural areas in 2040, fast internet access is considered crucial to support their development. The expansion of digitization and digital technologies is viewed as one of the prime needs of rural areas. Various authors (Whitacre and Manlove, 2016; Nah et al., 2021) have underlined the influence of emerging technologies and media (internet, social and mobile media) on civic participation in rural communities and the need for widespread fast access to attain a higher level of participation and information. The area covered by the project does indeed have the above needs, brought to light by the pandemic in an even more disruptive way, emphasizing the gap in internet connection quality between urban and rural areas.

Finally, the absence of replicability and continuity over time and the failed transition from isolated interventions to systemic actions remain among the most important critical points of social innovation projects in rural areas (Chiodo and Salvatore, 2021). This is due to a lack of continuity both in the funding programs and in the projects supported once the programs end and to difficulties in 'institutionalization', i.e., the taking charge by the local administrations of the interventions and their inclusion into ordinary local development policies (Ibidem). The coherence of the objectives of the CDC project with rural development policies and the initiatives envisaged by the new PNRR represent an opportunity for continuity, but the local planning capacity for accessing funding tenders still seems limited, so the process started by the project appears to need additional external support.

References

Adler, R.P., & Goggin, J. (2005). What Do We Mean By "Civic Engagement"?. ***Journal of Transformative Education, 3***(3), 236-253. doi:10.1177/1541344605276792

Agnoletti, M. (2014). Rural landscape, nature conservation and culture: Some notes on research trends and management approaches from a (southern) European perspective. ***Landscape and Urban Planning, 126***, 66-73. doi:10.1016/j.landurbplan.2014.02.012

Albanesi, C. (2005). ***I Focus Group***, Rome: Carocci.

Anderson, C.D., & Bell, M.M. (2003). The devil of social capital: a dilemma of American rural sociology. In P. Cloke (Ed.), ***Country Visions*** (pp. 232-244). Harlow, Essex: Pearson Education.

Arzeni, A., & Storti, D. (2021). L'innovazione come leva dello sviluppo territoriale: riflessioni sul contributo della Strategia Nazionale Aree Interne (SNAI). ***Agriregionieuropa, 4***. Available at https://agriregionieuropa.univpm.it/it/content/article/31/60/linnovazione-come-leva-dello-sviluppo-territoriale-riflessioni-sul-contributo

Barke, M., & Newton, M. (1997), The EU LEADER Initiative and endogenous rural development: The application of the programme in two rural areas of Andalusia, Southern Spain. ***Journal of Rural Studies, 13***, 319-341. doi:10.1016/S0743-0167(97)00027-2

Bobbio, L. (2004). ***A più voci. Amministrazioni pubbliche, imprese, associazioni e cittadini nei processi decisionali inclusivi, Cantieri Analisi e strumenti per l'innovazione***. Naples: Edizioni Scientifiche Italiane.

Bock, B.B. (2015). Rural marginalisation and the role of social innovation; a turn towards nexogenous development and rural reconnection. ***Sociologia Ruralis, 56***(4), 552-573. doi:10.1111/soru.12119

Bosworth, G., Annibal, I., Terry, C., Price, L., Sellick, J., & Shepherd, J. (2016). Empowering local action through neo-endogenous development; the case of LEADER in England. ***Sociologia Ruralis, 56***(3), 427-449.

doi:10.1111/soru.12089

Bryman, A. (2008). ***Social Research Methods***. New York: OU Press.

Butterfoss, F.D. (2006). Process Evaluation for Community participation. ***Annual Review of Public Health, 27***, 323-34. doi:10.1146/annurev.publhealth.27.021405.102207

Charmaz, K. (2006). ***Constructing Grounded Theory: a Practical Guide through Qualitative Analysis***. London: Sage Publications.

Chartier, O., Salle, E., Irvine, K., Kull, M., Miller, D., Nieto, E., Vestergård, L.O., Potters, J., Slätmo, E., Zomer, B., & Iadecola, F. (2021). ***Long-Term Vision for Rural Areas: Contribution from SHERPA science-society-policy platforms***. SHERPA Position Paper. doi:10.5281/zenodo.4557440

Checkoway, B., & Gutiérrez, L. (2006). Youth participation and community change: An introduction. ***Journal of Community Practice, 14***(1/2), 1-9. doi:10.1300/J125v14n01_01

Chiodo, E., & Salvatore, R. (2021). Repertorio di buone pratiche per l'agricoltura delle aree interne. ***Agriregionieuropa, 3***. Available at https://agriregionieuropa.univpm.it/it/content/article/31/59/repertorio-di-buone-pratiche-lagricoltura-delle-aree-interne

Corbin, J., & Strauss, A. (2015). ***Basics of qualitative research: techniques and procedures for developing grounded theory***. Thousand Oaks, Los Angeles, CA: Sage Publications.

Corrao, S. (2005). ***Il Focus Group***. Milan: FrancoAngeli.

Cronbach, L.J. (1980). ***Toward Reform of Program Evaluation: Aims, Methods, and Institutional Arrangements***. San Francisco, CA: Jossey-Bass Publisher.

Curtis, A., Ross, H., Marshall, G.R., Baldwin, C., Cavaye, J., Freeman, C., Carr, A., & Syme, G.J. (2014). The great experiment with devolved NRM governance: lessons from community engagement in Australia and New Zealand since the 1980s. ***Australasian Journal of Environmental Management, 21***(2), 175-199. doi:10.1080/14486563.2014.935747

De Rita, G., & Bonomi, A. (1998). ***Manifesto per lo sviluppo locale***. Turin: Bollati Boringhieri.

Díaz-Puente, J.M., Montero, A.C., & de los Ríos Carmenado, I. (2007). Empowering communities through evaluation: some lessons from rural Spain. ***Community Development Journal, 44***(1), 53-67. doi:10.1093/cdj/bsm008

Elder, G.H., & Conger, R.D. (2000). ***Children of the land: Adversity and success in rural America***. Chicago, IL: University of Chicago Press.

Féret, S., Berchoux, T., Requier, M., Abdelhakim, T., Slatmo, E., Chartier, O., Nieto, E., & Miller. D.R. (2020). ***Framework Providing Definitions, Review and Operational Typology of Rural Areas in Europe***. Deliverable D3.2 of the SHERPA Sustainable Hub to Engage into Rural Policies with Actors Project. Available at https://rural-interfaces.eu/ wp-content/uploads/2020/04/SHERPA_D3-2_Framework-rural-definition.pdf

Fiorina, M.P. (1999). Extreme voices: a dark side of civic engagement. In T. Skocpol & M.P. Fiorina (Eds.), ***Civic Engagement in American Democracy*** (pp. 395-425). Washington, DC: Brookings Institute Press.

Fondazione Compagnia di Sanpaolo. (2018). ***Progetti selezionati CivICa 2018***. Available at https://www.compagniadisanpaolo.it/wp-content/uploads/Booklet-CivICa-2018.pdf

Fondazione Compagnia di Sanpaolo. (2021). ***CivICa. Progetti di Cultura e Innovazione Civica***. Available at https://www.compagniadisanpaolo.it/it/progetti/civica-progetti-di-cultura-e-innovazione-civica/

Godwin T.A., Martin, K.E., Gasteyer, S.P., Keating K., & Pigg, K. (2012). Community leadership development education: promoting civic engagement through human and social capital. ***Community Development, 43***(1), 31-48. doi:10.1080/15575330.2011.645043

Hardin, G.J. (1968). The Tragedy of the Commons: The population problem has no technical solution; it requires a fundamental extension in morality. ***Science, 162***(3859), 1243-1248. Available at https://www.science.org/doi/10.1126/science.162.3859.1243

Henry, G.T., & Mark, M.M. (2003). Understanding evaluation's influence on attitudes and actions. ***American Journal of Evaluation, 24***, 293-314. doi:10.1016/S1098-2140(03)00056-0

Horlings, L. (2015). The worldview and symbolic dimensions in territorialisation: how human values play a role in a Dutch neighbourhood. In J. Dessein, E. Battaglini & L. Horlings (Eds.). ***Cultural Sustainability and Regional Development*** (pp. 43-58). London and New York: Routledge.

ISPRA (Istituto Superiore per la Protezione e la Ricerca Ambientale). (2018). ***Rapporto nazionale pesticidi nelle acque - dati 2015-2016. Edizione 2018***. Rome: ISPRA. Available at https://www.isprambiente.gov.it/files2018/pubblicazioni/rapporti/Rapporto_282_2018.pdf

Johnson, K. (2006). ***Demographic trends in rural and small town America***. Durham, NH: University of New Hampshire, Carsey Institute.

Kleiner, T.M. (2020). The isolating side effect of civic participation. ***Journal of Sociology, 57***(4), 877-897. doi:10.1177/1440783320969906

Kleiner, T.M. (2021). Civic Participation and Social Embeddedness: Differences Between Urban and Rural Communities. ***The International Journal of Community and Social Development***, *3*(1), 45-67. doi:10.1177/25166026211002048

Kvale, S. (2007). ***Doing Interviews***. London: Sage Publications.

Lay, J.C. (2006). Learning about politics in low-income communities: Poverty and political knowledge. ***American Politics Research, 34***(3), 319-340. doi:10.1177/1532673X05278

Li, Y., Westlund, H., Zheng, X., & Liu, Y. (2016). Bottom-up initiatives and revival in the face of rural decline: Case studies from China and Sweden. ***Journal of Rural Studies, 47***, 506-513. doi:10.1016/j.jrurstud.2016.07.004

Li, Y., Westlund, H., & Liu, Y. (2019). Why some rural areas decline while some others not: An overview of rural evolution in the world. ***Journal of Rural Studies, 68***, 135-143. doi:10.1016/j.jrurstud.2019.03.003

Lowndes, V., & Wilson, D. (2001). Social capital and local governance: exploring the institutional design variable. ***Political Studies, 49***, 629-647. doi:10.1111/1467-9248.003

Lumino, R. (2013). ***Valutazione e teorie del cambiamento.*** Milan: FrancoAngeli.

Magnaghi, A. (2000). ***Il progetto locale.*** Turin: Bollati Boringhieri.

Meyer, J. (2000). Evaluating action research. ***Age Ageing, 29***(2), 8-10. doi:10.1093/oxfordjournals.ageing.a008104

Midgley, G. (2011). Theoretical pluralism in systemic action research. ***Systemic Practice and Action Research, 24***(1), 1-15. doi:10.1007/s11213-010-9176-2

Milone, P., & Ventura, F. (2019). New generation farmers: Rediscovering the peasantry. ***Journal of Rural Studies, 65***, 43-52. doi:10.1016/j.jrurstud.2018.12.009

Nah, S., Kwon, H.K., Liu, W., & McNealy, J.E. (2021). Communication Infrastructure, Social Media, and Civic Participation across Geographically Diverse Communities in the United States. ***Communication Studies, 4***(3), 437-455. doi:10.1080/10510974.2021.1876129

Niessen, B. (2018). ***La storia del 'Noi': cambiare la cultura per aprire gli spazi di cittadinanza.*** Turin: Fondazione Compagnia di Sanpaolo. Available at https://www.compagniadisanpaolo.it/wp-content/uploads/1-La-storia-del-Noi-di-Bertram-Niessen.pdf

Noguchi-Shinohara, M., Hirako, K., Tsujiguchi, H., Itatani, T., Yanagihara, K., Samuta, H., & Nakamura, H. (2020). Residents living in communities with higher civic participation report higher self-rated health. ***PLoS ONE, 15***(10). doi:10.1371/journal.pone.0241221

Ostrom, E. (1990). ***Governing the Commons The Evolution of Institutions for Collective Action***. Cambridge: Cambridge University Press.

Pawson, R., & Tilley, N. (1997). ***Realistic evaluation***. London: Sage Publications.

Perna, T. (2002). ***Aspromonte I parchi nazionali nello sviluppo locale.*** Turin: Bollati Boringhieri.

Portales, L. (2019). ***Social Innovation and Social Entrepreneurship Fundamentals, Concepts, and Tools.*** Cham, Switzerland: Palgrave Macmillan. doi:10.1007/978-3-030-13456-3

Price, M., Miller, D.R., McKeen, M., Slee, B., & Nijnik, M. (2017). ***Categorisation of Marginalised Rural Areas (MRAs).*** Deliverable 3.1. Social Innovation in Marginalised Rural Areas (SIMRA). Report to the European Commission. doi:10.5281/zenodo.3625493

Pritzker, S., & Metzger, A. (2011). Facilitating Civic Engagement Among Rural Youth: A Role for Social Workers. In L.H. Ginsberg (Ed.), ***Social work in rural communities*** (pp. 185-209). Alexandria, VA: Council on Social Work Education.

Putnam, R. (1993a). ***Making Democracy Work: Civic Traditions in Modern Italy***. Princeton, NJ: Princeton University Press.

Putnam, R. (1993b). The Prosperous Community: Social Capital and Public Life. ***The American Prospect, 4***(13), 35-42.

Putnam, R. (2000). ***Bowling Alone – the Collapse and Revival of American community***. New York: Touchstone Books.

Rapley, T. (2007). ***Doing Conversation, Discourse and Document Analysis***. London: Sage Publications.

Ray, C. (2000). The EU LEADER programme: rural development laboratory. ***Sociologia Ruralis, 40***(2), 163-172. doi:10.1111/1467-9523.00138

Roßteutscher, S. (Ed.). (2005). ***Democracy and the Role of Associations. Political, Structural and Social Contexts.*** London and New York: Routledge.

Rossi, P.H., Lipsey, M.W., & Freeman, H.E. (2004). ***Evaluation: A systematic approach***. Thousand Oaks, CA: Sage Publications.

Ryan, V.D., Agnitsch, K.A., Zhao, L., & Mullick, R. (2005). Making sense of voluntary participation: A theoretical synthesis. ***Rural Sociology, 70***(3), 287-313. doi:10.1526/0036011054831198

Salvatore, R., & Chiodo, E. (2017). ***Non più e non ancora. Le aree fragili tra conservazione ambientale, cambiamento sociale e sviluppo turistico.*** Rome: FrancoAngeli.

Sanchez-Zamora, P., Gallardo-Cobos, R., & Ceña-Delgado, F. (2014). Rural areas face the economic crisis: Analyzing the determinants of successful territorial dynamics. ***Journal of Rural Studies, 35,*** 11-25. doi:10.1016/j.jrurstud.2014.03.007

Schön, D.A. (2017). ***The Reflective Practitioner How Professionals Think in Action***. London and New York: Routledge.

Scriven, M. (1996). Types of Evaluation and Types of Evaluator. ***American Journal of Evaluation, 17***(2), 151-161. doi:10.1177/109821409601700207

Shortall, S. (2008). Are rural development programmes socially inclusive? Social inclusion, civic engagement, participation, and social capital: Exploring the differences. ***Journal of Rural Studies 24***(4), 450-457. doi:10.1016/j.jrurstud.2008.01.001

Stame, N. (2004). Theory-based Evaluation and Types of Complexity. ***Evaluation, 10***(1), 58-76. doi:10.1177/1356389004043135

Starosta, P. (2010). Civic Participation in Rural Europe. ***Przegląd Socjologiczny, 59***, 77-108. Available at https://www.ceeol.com/search/article-detail?id=174401

United Nations. (1992). ***Agenda 21***. Available at https://www.un.org/esa/dsd/agenda21/

United Nations. (2015). ***Transforming our world: the 2030 Agenda for Sustainable Development***, Resolution adopted by the General Assembly on 25 September 2015. Available at https://www.un.org/ga/search/view_doc.asp?symbol=A/RES/70/1&Lang=E (17/01/2021)

Van der Ploeg, J.D. (2008). ***I nuovi contadini: le campagne e le risposte alla globalizzazione***. Rome: Donzelli Editore.

Vercher, N. (2022). The Role of Actors in Social Innovation in Rural Areas. ***Land, 11***(5), 710. doi:10.3390/land11050710

Warren, M.E. (2001). ***Democracy and association***. Princeton, NJ: Princeton University Press.

Weiss, C. (1997). Theory-Based Evaluation: Past, Present, and Future. ***New Directions for Evaluation, 76***, 41-55. doi:10.1002/ev.1086

Weiss, C. (2007). La valutazione basata sulla teoria: passato, presente e futuro. In N. Stame (Ed.) ***Classici della Valutazione***. Milan: FrancoAngeli.

Whitacre, B.E., & Manlove, J.L. (2016). Broadband and civic engagement in rural areas: What matters?. ***Community Development, 47***(5), 700-717. doi:10.1080/15575330.2016.1212910

Wieliczko, B., Kurdyś-Kujawska, A., & Floriańczyk, Z. (2021). EU Rural Policy's Capacity to Facilitate a Just Sustainability Transition of the Rural Areas. ***Energies, 14***(5050). doi:https://doi.org/10.3390/en14165050

Wilcox, D. (2003). ***The guide to effective participation***. Available at http://www.partnerships.org.uk/guide

Za, L. (2001). ***Comunità memoria sviluppo***. Lecce: Pensa Multimedia.

Zammuner, V. (2003). ***I Focus Group***. Bologna: Il Mulino.

Author Contributions: E.P.: conceptualization, methodology, investigation, analysis, writing, review and editing of the manuscript, supervision; B.L.P. and G.A.: funding acquisition, analysis of the results. All authors have read and agreed to the published version of the manuscript.

Funding: This research was funded by the Compagnia di Sanpaolo Foundation within the framework of the CivICa 2019 call.

Acknowledgements: Very special thanks to all the members of the local communities of Rovasenda and Buronzo, who spared no time and effort to act as concerned citizens and greatly contributed to the project.

Chapter 7

Young (K-5) Environmental Agents Course

Importance of remote practical elementary school environmental education at a rural school in the state of Rio de Janeiro, Brazil

Elvio Kamiyama[1], Nathallia Mercedes Miranda[2], and Flavia Aline Souza Brasileiro[3]

Abstract

During the pandemic in 2021, the online course: "Young (K-5) Environmental Agents", an early childhood environmental education course, was administered online to a rural public school in Rio de Janeiro, Brazil. In total, 58 students were involved with the aim of teaching the principles of biomes through practical and online activities. The course consisted of carrying out several practical environmental education activities over a period of 3 months that students carried out in their homes and amid their immediate natural surroundings in order to create greater awareness and sensitization to the core theme of the course. The syllabus included theoretical explanatory online classes complemented by streaming videos, theoretical practical activities using delivered course materials, and common materials the children had in their homes and collected in nature. Before completing the course, students were required to do a final project and give an online presentation. The pedagogical staff carried out final evaluations. The results supported positive participation assessments for children and their parents, raising children's awareness of the proposed themes. Participation by students and parents in the theoretical component was highly satisfactory, while participation in practical exercises was nearly 100% among students. All activities were voluntary. The project documents the importance of practical activities for effective environ-

1 Social Assistance Program – Social Service of Commerce RJ
2 Social Assistance Program – Social Service of Commerce RJ
3 Social Assistance Program – Social Service of Commerce RJ

mental education, the utility of early online education, and the importance of the child's contact with its immediate natural surroundings.

Keywords: K-5 Environmental On-line Education, Practical Environmental Activities, Reflective Education

Introduction

The Young Environmental Agents course was administered during the second quarter of 2021. Due to the COVID-19 pandemic and public health restrictions in the State of Rio de Janeiro, Brazil, a distance-learning platform was used to pilot this early childhood environmental education course at a rural public school during the first cycle of elementary education at that time. The course is the fruit of the SESC+ Sustainability Project, an initiative of the Social Service of Commerce of the State of Rio de Janeiro (SESC RJ) and is within the preview of sustainability initiatives for the external public. As part of its mission, SESC RJ promotes initiatives that disseminate knowledge, inspire people and strengthen networks within communities and wider territories. SESC RJ is committed to sustainability and community development through the promotion of educational actions in schools, social institutions and companies. In addition to offering quality content to audiences of all ages, it offers courses, workshops, lectures, debates, conversation circles, exhibitions, guided visits and experiences. The course Young (K-5) Environmental Agents was drawn from the epistemology developed by the Brazilian educator Paulo Freire (2000), which engages the student with experiential activities that foster curiosity and questioning about the immediate surrounding reality. Thus, this project generates learning in environmental education through practical activities that transform a student's thinking about the world.

The heightened isolation generated in a rural community during the pandemic provided both the developmental need and opportunity for young elementary school children (K-5) to remain engaged in their educational experience. Young Environmental Agents was designed to generate moments of socio-environmental contact that could enrich their knowledge beyond the immediate overtaxed pandemic home environment with the idea of exploring the ecologies of their garden and street, which, according to Freire's theory, are viable alternatives to generate a change of thinking in the students.

One concept used in the methodology was the inverted classroom, whereby the children assumed the responsibility to work at their own pace offline, while the online dialectic synthesis focused on the results obtained in each class (Blikstein, 2013). The online classes, which administered the lessons to be carried out at home, were always accompanied by parents, who showed extreme interest in the proposed activities that required common household materials.

The course methodology assumed a human mental condition characterized as being unfinished, whereby an individual becomes an agent capable of daily transformation, both internally and outwardly, in the environment in which one lives (Loureiro & Torres, 2014). The ensuing results are not only concretely visible but also linguistically and symbolically changed in the children's speech about the environment in which they live.

Course content pertained to the fundamentals of biomes and field biodiversity, divided into three topics: soil, water, and garden. As a school and community demand-based course, all development and planning were carried out in a participatory manner with principal stakeholders. The theme is extremely important for future generations, as Brazil is home to around 15% of the world's biodiversity. In particular, the state of Rio de Janeiro is located in the "Mata Atlântica" (Atlantic forest) biome, which, despite having originally been one of

Figure 1. Young Environmental Agents On-line activities Source: the authors (2021).

the richest areas in terms of biodiversity in the world, has already lost about 93% of its original area due to deforestation and urbanization. Although it is one of the most threatened areas on the planet in terms of loss of biodiversity, the Mata Atlântica biome is home to a wide variety of endemic species. It is considered a priority "Hot Spot" for the conservation of global biodiversity (Fundação SOS Mata Atlântica, 2023).

Research questions / purpose of the program

The Young Environmental Agents program aims to:

- Offer students a differentiated education based on excellence in environmental education, in accordance with SDG 4 – Quality Education (UN, 2022);
- Teach and promote a transformative and reflective education about the actions children can take to deal with local changes. Encourage students to take actions that contribute to environmental conservation and consequently reduce climate change, by SDG 13 - Action against global climate change (UN, 2022);
- Sensitize students through practical environmental education activities that turn them into agents of change in their environment;
- Use methodology that emphasizes reflection for change, giving the child the leading role in the applied lessons;
- As the program took place during the pandemic, it was essential to prioritize the surroundings of the environment in which they live, such as their garden, house, street, or neighborhood.

Methodology / program description

The project was developed in participation with the Marquês de Salamanca Municipal School faculty, a rural public school in the city of Três Rios, in the State of Rio de Janeiro, Brazil. The target audience was K-5 classrooms comprised of 58 low-income students. The 3-month course (April to June 2021) included 21 hours of remote, theoretical classes, 23 hours of offline practical activities, and 3 hours of presentation (1 hour for each sub-thematic module).

The overall themes of the course were biomes and field biodiversity. The theoretical classes included a complementary practical explanation of the proposed practical activities. The students used didactic material delivered by SESC to the school, which was then distributed to the students. In addition, students were responsible for procuring common household items in their homes and collecting biological assets from nature found in their gardens. In order to receive a certificate of completion, students were required to construct a model of the local biodiversity and give an online presentation.

Inverted classroom activities included:

- Crop burning wildfires: As the region suffers annual crop burning, the clearing of straw left after harvest, a traditional albeit debilitating agricultural practice, has destroyed large swathes of forest through wildfires. Though popularly justified by tradition, crop burning has hampered conservation efforts and has jeopardized local biodiversity. Being a high-priority environmental topic, the practical activity was to make a poster that expresses your feelings regarding this practice;
- Biodiversity in the garden: Understanding one's natural surroundings and the ecology of the family garden was a key objective of the course. All students learned how to make and maintain a mini terrarium;
- Pollinators: Who, What, Where and Why of important insect pollinators. Practical activity and origami modeling of some insects;
- Water use;
- Water biodiversity: Students created models drawn from observations of what the students perceived as aquatic biodiversity in the bodies of water and water courses close to them;
- Soil types and degradation: Erosion due to soil misuse is paramount in this mountainous region. Young Environmental Agents had a practical activity explaining different soil formations;
- Taxonomy of leaves: Students collected leaves from trees and plants in their garden first to observe the different species;
- Identification and exsiccation of leave, the production of dry leaves: this practical experience fostered student's interest in scientific nomenclature;
- Riparian forests explained;
- "Conservation begins with us!": raising awareness about how

global concern for the environment starts in the homes, our schools, neighborhoods and cities, etc.;
- Final Project: each student builds a model that represents his or her perception of the local biodiversity;
- Final On-line Presentations.

Course methodology drew significantly from the teachings of Paulo Freire, an internationally acclaimed Brazilian pedagogue who readily uses the natural environment as a learning platform. The underlying objective of this method is to allow students to reflect on their experience to foster critical thinking, which, in turn, transforms one's reality. Through practice and engagement in concrete actions, agency for action accumulated individually and was aggregated collectively. Thus, young children become environmental agents for change. (Freire, 2016).

For example, in the forest fire module, students had to observe their local environment and reflect on the cultural practice of burning to clear pasture. How does this practice affect the biodiversity that surrounds them? Do alternative concrete and attainable actions exist? Since parents are mainly responsible for the burnings, sensitization of the parents of viable options by the children (agency) could catalyze a change in the current collective reality.

Children are drawn into an interactive form of environmental education by observing their surroundings and carrying out a critical analysis of their home, garden and neighborhood. The reflection component of each exercise, that is, the antithesis of the existing problem, is anchored to how one should relate to their environment in a sustainable way (Orr, 1992) while always being a practical, playful and fun activity.

Another way of engaging students and making classes more productive was the intermittent use of the flipped classroom model, in which some lessons were previously sent by streaming video and then discussed in online classes.

It is worth reiterating that this course took place at the height of the pandemic and that the school was located 12 km from the nearest paved road. Thus, all activities were online, which required experimentation and adaptation of the lessons by the teachers, school director, parents, students and course instructor. Used excellently and

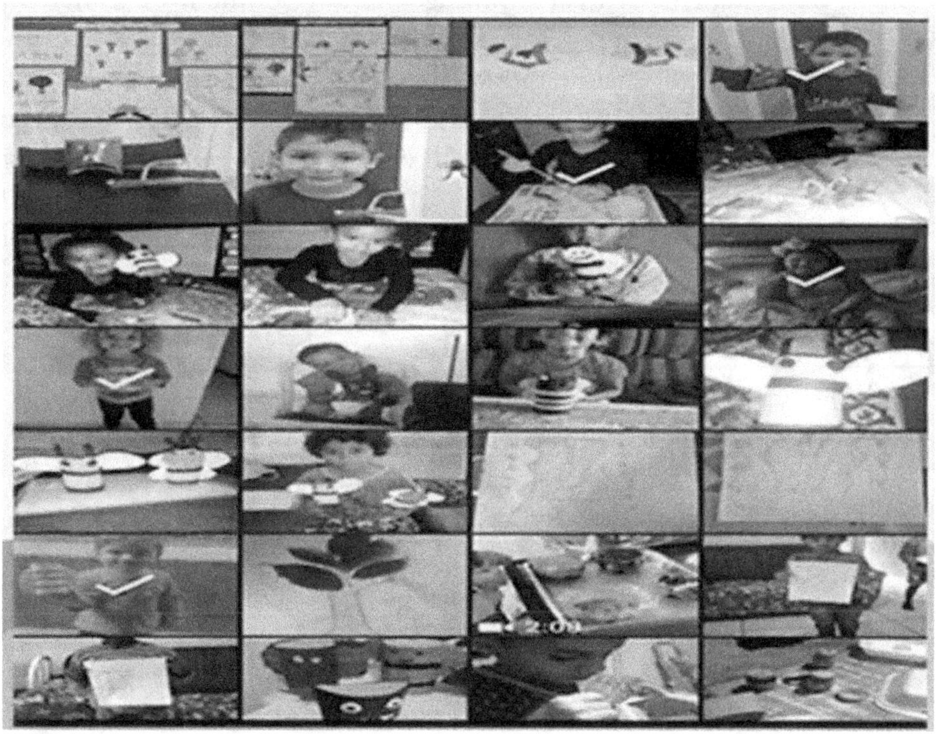

Figure 2. Pollinator online activities Source: the authors (2021).

productively, the activities generated a significant satisfactory amount of completed work and class participation.

Literature review

Paulo Freire, an important Brazilian author and educator, stands out not only for changing the way of thinking about education but also for restructuring how students are trained, making them critical and active actors in social and political change (Freire, 1970). His internationally recognized treatise, *Pedagogy of the Oppressed*, transformed how marginalized populations were educated. Less known is that Freire puts the student-environment dialectic at the center of his model, whereby students are responsible for being active in changes that improve our place of residence, community, and school, promoting solutions that the students themselves can help resolve environmental issues.

The Young (K-5) Environmental Agents program methodology included other Freire pedagogies:

- The Pedagogy of Autonomy (Freire, 1987) is based on the development of critical thinking, the development of leadership skills, the encouragement of personal responsibility and the promotion of sustainability. Thus, the student becomes responsible, autonomous, critical, and an agent of change around him, with sustainable thoughts and real concern for the environment in which he or she lives;

- The Pedagogy of Hope (Freire, 1996) is a pedagogy based on transforming actions, community initiatives, and students as agents of social and environmental change;

And finally,

- The Pedagogy of Solidarity (Freire, 1998), in which the author encourages actions that can collaborate with the social environment in which they live. The feeling of helping others, students develop the necessary feelings for socio-environmental thinking for sustainability.

This course is geared toward local change, not only theoretically but also fostering change in how children think and how they can be influencers and multipliers for changing their school, neighborhood, street, and where they live. This thought is in line with the idea of Costa (2005), who advocates that environmental education should be capable of promoting the formation of critical and aware subjects who are capable of understanding the social, cultural and economic underpinnings of environmental issues and willing to seek solutions for social and environmental transformation.

The Young (K-5) Environmental Agents course emphasizes that actions should not be just theoretical. It focuses on practice and concrete actions based on the reality in which students live. Thus, since its inception in 2019 at SESC-RJ, the intention has been to train "Young Environmental Agents" who understand the rudiments of environmental stewardship and sustainability. Thus, based on a series of participative discussions with local stakeholders, teachers, management, students, and community members, training and practical action became the course's foundation. For example, for the biodiversity of the field module, students examine the local challenges, and then, through practical actions, students consolidate learning.

Stakeholders agreed to keep environmental education content relevant to the local environment. Theoretically, Orr (1992) emphasizes the importance of connecting with the students' surroundings and using real actions that pertain to their environment. Microenvironments are complex systems, and environmental education must strive to elucidate an integrated understanding of the whole that surrounds students. In this way, young citizens become aware and capable of making necessary social and environmental changes. Smith (2012) advocated for a contextualized environmental education adapted to students' immediate needs and interests by taking into account students' cultural and socioeconomic perspectives and involving local communities in the environmental education process, agency for percolates throughout the community.

The biomes and species module is extremely relevant, as Brazil possesses the most biodiversity in the world with six different biomes: the Atlantic Forest, Amazon rainforest, Cerrado (highland savanna), Caatinga (xeric shrubland), Pantanal (tropical wetland) and the Coastal and Marine System. Rich in fauna and flora diversity, especially in innumerable endemic species, the biodiversity of these Brazilian biomes is staggering. Combined, they contain more than 20% of all species of animals and plants in the world: 103,000 animal species and 43,000 (and counting) plant species, making Brazil a world hotspot for biodiversity conservation (May 2019).

The Global Assessment Report on Biodiversity and Ecosystem Services, published in 2019 by the Intergovernmental Platform on Biodiversity and Ecosystem Services (IPBES), identified five main causes of biodiversity loss:

- Changes in public policy on land use and land cover, including conversion of natural ecosystems into agricultural or urban areas;
- Over- and unsustainable exploitation of natural resources, including overfishing, over-logging and illegal hunting;
- Climate change: global warming and the associated extreme weather events;
- Pollution: air, water and soil pollution, especially chemical and plastic pollution;
- Invasive species that threaten native species and disrupt ecosystems.

These causes are interconnected and mutually reinforcing, thereby increasing pressure on biodiversity and straining ecosystem services. The IPBES report highlights these negative trends and the economic, social and political factors driving them, including population growth, unsustainable consumption and lack of consideration for biodiversity and ecosystem services in decision-making and public policy (United Nations, 2019).

In Brazil, according to MAPBIOMAS (2020), a survey carried out from 1985 to 2020 dramatically confirms that each year, Brazil burned an area larger than that of England (150,957 km^2 per year). During this 36-year window, Brazil burned 1,672,142 km^2, the equivalent of 19.6% of the national territory. An incredible 65% of the total area burned was native vegetation!

Among the necessary actions called for the conservation of biodiversity (IPBES, 2019) are strengthening and effective application of environmental policies and laws, integrating biodiversity into other public policies and sectors, ecological restoration, strengthening protected areas and indigenous and community territories, promoting sustainable production and consumption practices and investing in research and monitoring.

This same report specifically indicates environmental education as a necessary intervention in mitigating the unrelenting global assault on biodiversity with these recommendations:

- Integration of environmental education at all levels of education, from early childhood education to higher education, to raise awareness of the importance of biodiversity and ecosystem services;
- Encourage informal education and lifelong learning: In addition to formal education, it is important to encourage informal environmental education and lifelong learning through initiatives such as community education programs and visits to parks and nature reserves, among others;
- Strengthening the capacity of educators and facilitators: It is important to strengthen the capacity of environmental education educators and facilitators so that they can convey accurate and up-to-date information about biodiversity and ecosystem services;
- Promotion of participatory and multidisciplinary approaches:

> Participatory and multidisciplinary approaches are vital to environmental education to involve local communities, indigenous and traditional populations, scientists and other relevant actors in biodiversity management.

The report emphasizes that promoting environmental education is essential to ensure people's awareness and engagement in biodiversity conservation, placing it a high priority in conservation policies and programs.

The Brazilian Atlantic Forest, known as the Mata Atlântica biome, encompasses the entire state of Rio de Janeiro and 15% of the Brazilian national territory. Although just 12.4% of its original coverage remains (Fundação SOS Mata Atlântica & Instituto Nacional de Pesquisas Espaciais, 2021), it possesses incomparable animal and plant diversity with many species endemic to the biome. Tragically, due to human activity through the expansion of agriculture, urbanization, deforestation, etc., the annual loss of fauna and flora of the Atlantic Forest biome supersedes all current conservation and preservation efforts to maintain the biodiversity within this ecosystem.

Regarding fauna, the Atlantic Forest biome is home to many endangered species of storybook notoriety, such as the jaguar, the ocelot, the golden lion tamarin and the muriqui, the largest primate in the Americas. In addition, the region remains unparalleled in birds, reptiles, amphibians and insects; many species have not yet been completely studied and cataloged.

The Atlantic Forest biome is home to a wide variety of plant species, especially species endemic to the biome. Among the best-known species are Brazilwood (Pau Brazil), the Fig Tree (Figueira), the Jequitibá (with some specimens over 3,000 years old), the Araucaria (a pine tree with a distinctive radial crown) and the Juçara Palm Tree (source of heart-of-palm), which is an endangered species.

Incidentally, and in no small manner of speaking, the Atlantic Forest is also home to roughly 70% of the Brazilian population (Ribeiro, 2009). Since its discovery and subsequent colonization by the Portuguese, human activity has heavily affected and continues to alter this biome to extinction. Nearly 93% of its original cover has been lost due to deforestation, urbanization, agricultural expansion and mining. Preserving Atlantic Forest biome biodiversity is essential for conserving Brazilian fauna and flora, as well as preventing the ecosystem services it provides, such as regulation of climate and water resources from falling into collapse.

Thus, early (K-5) environmental education on Atlantic Forest biodiversity is instrumental to the children's future and their communities that call the forest home. Reinforcing and forming children capable of thinking critically and acting concretely in their own "backyard"

Figure 3. Practical activities - Exsiccate. Source: the authors (2021).

is a pivotal investment in the future survival of the other 200 million Brazilians who unwittingly draw on the biodiversity resources of the Mata Atlântica forest.

Results

The results were commendable, not only because of the activities developed in the participation of local stakeholders but also because student participation was always above 70%. Remembering that the children resided in a remote rural location with only online access during the pandemic. They were very young learners divided into early childhood education and two classes of the first cycle of elementary school.

The project kicked off with a planning session in April with the SESC-Três Rios instructor and analyst and the Director of the

Marques de Salamanca Municipal School. This planning session organized the dates and first triage of proposed topics. The Young (K-5) Environmental Agents" project was presented to students and teachers using the Google Meet platform shortly afterward. With the parents and children onboard, the inaugural class of the Forest Fire and Biodiversity module went online. Students and parents participated avidly with questions and anecdotal reports of local fires and fire as an agricultural practice. This joint session with the parents reinforced the underlying pedagogic strategy of heightening the perception of the immediate surroundings (Freire, 1996) and hands-on learning that can facilitate change in their current reality.

In synthesizing the work presented on forest fire, both the children and their parents were sensitized to the fact that crop-burning wildfires are one of the greatest causes of loss of biodiversity, even though it is a culturally accepted annual practice in the region for cleaning pasture for local agriculture or sometimes just for cleaning an area. The children talked up the topic with enthusiasm and concern as they began to put this practice into a new perspective. Their thoughts changed as they learned and sought ways to transform the agricultural practices of the adults without the use of forest fires.

The month of May saw the full rollout of the program. The classes continued for each group with a different thematic subject within the proposed one. The pedagogic approach consisted of the biodiversity of gardens and orchards for kindergarten students, water biodiversity from 1st to 3rd grade, and soil biodiversity and its use for 4th and 5th graders. The local reality was constantly used as the underlying platform to discover how to improve upon what already existed in the community. The course presented work and activities from projects previously implemented by institutions and non-governmental organizations, such as TAMAR, the Marine Turtle Project, a highly successful Brazilian conservation project for marine species conservation. Also, several course activities were carried out with CEIVAP, the Committee for Integration of the Paraíba do Sul River Basin, an important river for Rio de Janeiro and Brazil. The students were interested and interacted readily with their parents, relating experiences, for example, about the Piabanha River and stories that their grandparents had told about how the river was in their youth. Thus, the children drew comparisons with the current situation.

Figure 4. Practical activities- types of soil. Source: the authors (2021).

The elementary school, represented by the manager, has a significant invested interest in developing projects that interact with the entire local community. According to one of the teachers, this project, administered by SESC-Três Rios, was of great importance for them to structure educational ideas in participation with the school, students and other responsible entities.

Student participation in the practical classes was a reliable indicator of positive student engagement in relation to learning the content offered. Regarding the practical activities, teachers encouraged students to use their creativity with reusable materials, such as toilet paper rolls, cardboard cores, PET bottles, aluminum cans, tree leaves and dry sticks, among others. It is extremely important to emphasize that hands-on activities stimulate emotional intelligence through the tactile, visual, auditory and olfactory senses to generate a more extensive perception of the environment and its ecological relationships (Milfont and Roder, 2010).

In June, the month in which World Environment Day is celebrated, themes emphasize how people can preserve the environment and, thus, maintain biodiversity. On June 5th, World Environment Day itself, children performed environmental monitoring tasks related to the United Nations 2030 Agenda for Sustainable Development (U.N., 2015). There are 17 Sustainable Development Goals (SDGs), which include SDG 6 - Safe water and sanitation, SDG 11 - Sustainable cities and communities and SDG 13

- Action against global climate change. Regular monitoring of the state of the environment is critical to achieving these goals (U.N., 2015).

In practical classes, the students produced drawings and collages that portrayed the local biodiversity from their backyard or schoolyard. Kindergarten-aged to 3-year-old children generated representative scenarios fashioned from reusable materials suggested in theoretical classes. There were four scenarios: a riparian forest, gardens with pollinators, marine environments of fish and other marine animals, and a river environment with fish and vegetation. Students from the 4th and 5th grades experimented with soil types with beans and grains, and in the final presentation, they performed an experiment that represented how the soil types behave during rainfall, which they presented during the final class with the participation of students, parents, professors, school management, and SESC-Três Rios.

As the practical progressed and students assimilated the content through experiential learning of culturally relevant themes, a new awareness by the students drove a desire to carry out extracurricular actions that could change their reality and local environment.

At the end of the course, the recently vaccinated SESC-Três Rios staff traveled more than 30 km to deliver medals and a certificate of course completion house to house to the children and families that had only been seen online. The children were euphoric, which was a

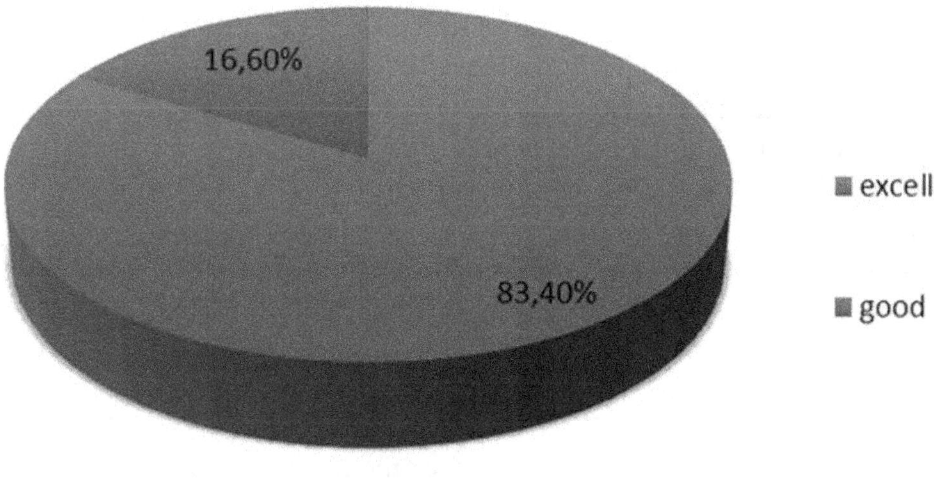

Chart 1. Final evaluation. Source: The authors (2021).

gesture that establishing the importance of "Young Environmental Agents" in their community was a success.

Conclusion & recommendations for theory and practice

It took three months (April to June 2021), 21 hours of remote and theoretical classes, 23 hours of practical activities, and 3 hours of presentation to prove the importance of practical activities in an effective environmental education program and the relevance of direct contact of the child with the environment in which he or she lives. Thus, Freire's educational models rang true: Providing an education that fosters independent critical thinking about one's surroundings and questioning what is in disagreement with correct ecological means is fundamental in the formation of agents of change. This was observed throughout the course, especially during the presentations and in the videos parents sent in on how the children were relating what was studied with what they were becoming conscious of in their community, for example, the annual fires, by making adults aware how they could be avoided, and if necessary making a formal complaint.

Smith's ecocentric future (2012) presents environmental education with the challenge of being critical and prepared for a more sustainable future. In this course, an ecocentric, pro-active posture that the future begins with small actions today, especially with very young children regarding topics such as the conservation of the soil rivers and care for the environment that surrounds us, was hardwired into this program since its inception.

According to Gusmão (2012), communities need a practical approach to environmental education, which allows students to have a deeper understanding of environmental problems and an active participation in solving these problems. The issues presented in this program raised both questions and suggestions for possible solutions from parents and students. The school volunteered to carry out activities for greater community involvement in addressing local environmental problems. In the same year that the school's sensory garden was created, the school vegetable garden was redesigned to include a repository for local history and environmental conservation.

In a follow-up to the course, a quality questionnaire was given to participants, and 83.40% responded excellent and 16.60% as good. From this, it can be understood that the course greatly benefited the students, so much so that in the second half of the same year, a continuation of the course on the biomes of Brazil was administered.

From birth to 12 years of age, childhood is the most important phase of cognitive development (Piaget, 1975). During this period, it is extremely important to allow children to explore and discover independently rather than passively receive information. Piaget believed that children need to actively interact with their environment to learn and that in doing so, they develop important skills such as critical thinking, problem-solving, and creativity. This was the direction of several lessons in the program; in addition to involvement with environmental issues, children were encouraged to reuse materials and explore their home environment for materials to carry out activities.

Despite the raging pandemic and innumerable restrictions on logistics, including fragile internet access, the results were impressive. In addition to the excitement of the children during the lively activities, the involvement of the community, parents, and the school in remote, poor rural regions was reassurance that the effort was a qualified success. It also observed through the activities and dialogues how much children are viscerally concerned about their future. The overall feeling was like the Young (K-5) Environmental Agents course was a life rope to an already compromised environmental future. The child dis-

Figure 5. Graduation of first Young Environmental Agents. Source: Authors (2021).

Figure 6. Children and selected activities. Source: Authors (2021).

played an empirical demand for involvement in the sustainability of the place where they live, their State and their families. This program can undoubtedly serve as a model for similar projects in the future.

References

Blikstein, P. (2013). *Maker-Centered Learning: Empowering Young People to Shape TheirWorlds.* John Wiley & Sons.

Costa, A. C. G (2005). *Educação Ambiental Crítica: nomes e endereçamentos da educação ambiental no Brasil* [Critical Environmental Education: Names and Approaches of Environmental Education in Brazil]. São Paulo: Cortez, 2005.

Freire, P. (1970). *Pedagogy of the Oppressed.* Rio de Janeiro: Paz e Terra.

Freire, P. (1987). *Pedagogy of Autonomy: Knowing Necessary for the Educational Practice.* São Paulo: Paz e Terra.

Freire, P. (1996). *Pedagogy of Hope: Reliving Pedagogy of the Oppressed.* Rio de Janeiro: Paz e Terra.

Freire, P. (1999). *Pedagogy of Solidarity.* Paz e Terra.

Freire, P. (2016). *Pedagogy of Indignation (Vol. 3).* Paz e Terra.Fundação SOS Mata Atlântica e Instituto Nacional de Pesquisas Espaciais. (2021). Atlas da Mata Atlântica: 2021. São Paulo: Fundação SOS Mata Atlântica.

Fundação SOS Mata Atlântica. (s.d.). *Mata Atlântica.* https://www.sosma.org.br/mata-atlantica/

Gusmão, P. H. L. de. (2012). *Práticas educativas de educação ambiental: reflexões eExperiências* [Educational Practices of Environmental Education: Reflections andExperiences]. Papirus Editora.

Loureiro, C. F. B., & Torres, J. R. (2014). *Educação ambiental: dialogando com Paulo Freire[Environmental Education: Dialoguing with Paulo Freire],* f. 92, p. 184.

Mapbiomas. (2020). *A cada ano, o Brasil queima* área *maior que a Inglaterra* [Every year, Brazilburns an area larger

than England]. https://mapbiomas.org/a-cada-ano-brasil-queima-areamaior-que-a-inglaterra
May, P. H. (2019). *Brazil's biodiversity: The largest in the world. Revista Brasileira deZoologia*, 36(1), 8-12. https://doi.org/10.1590/1678-4162-9583
Milfont, L., & Roder, S. (2010). *Nature as a model: 60 activities for environmental education.* Publisher Wak.
Orr, D. W. (1992). *Ecological literacy: Education and the transition to a postmodern world.*SUNY Press.
Piaget, J. (1975). *The construction of reality in children.* Editora Atica.
Smith, G. A. (2012). *Environmental education in the 21st century: Theory, practice, progress, andpromise.* Routledge.
United Nations. (2015). *Transforming our world: the 2030 Agenda for Sustainable Development.* https://www.un-.org/en/development/desa/population/migration/generalassembly/docs/globalcompact/A_RES_70_1_E.pdf
United Nations. (2019). *Global Assessment Report on Biodiversity and Ecosystem Services.*https://ipbes.net/global-assessment

Chapter 8

Assessment and Management of flood risks in the EU

Goals and Limitations depicted at the German river Murr

Andreas Brunold

Abstract

The floods in parts of Germany in July 2021 indicate that the ongoing anthropogenic climate crisis is leading to changes in the hydrological cycle. These deviations from a hydrologic balance lead to drought periods on the one hand and to heavy rainfall and flood events on the other. This case study concerns the Murr, a small German river with a high flood risk, and argues that the flood risk management of the EU and its member states has further to be constituted as a key environmental challenge of the 21st century.

To transfer European legislation into German legislation, the EU Directive 2007/60/EC on assessing and managing flood risks was turned into the National Water Act in 2010. Due to the federal structure of Germany, the law amendment in Baden-Württemberg, the federal state where the Murr is located, took place in 2013. Against this background, this case study on the River Murr examines two main research premises. Firstly, the study argues that the German federal structure leads to negative consequences for rivers with high flood risks. Secondly, it is to show that sustainable water resource management and processes of urban development and planning often are in diametrical oppositions due to economic interests.

The case study builds on data collected using a standardised questionnaire. The survey was conducted as a reaction to applying technical flood prevention measures in different villages and towns in the Murr area. The questionnaire was addressed to different kinds of experts in the field of flood prevention, such as water districts, lower water authorities and citizens' groups, among others.

The key findings of this case study are:

- Flood risk management without the participation of the public leads to more expensive and less sustainable technical flood prevention solutions.
- Flood risk management using technical flood prevention solutions such as flood protection walls are still commonly used in Germany, usually due to economic interests.
- Technical flood prevention solutions are usually deployed without any technology assessment processes.
- As meadows and riverside areas are still used as building grounds, technical flood prevention leads to an accentuation of the upstream and downstream riparian problems.
- In comparison to environmental associations and the citizens affected, there is a higher rate of acceptance for ecologically sustainable solutions, such as retention areas, than for technical solutions.

The paper concludes with three main points. The authors argue in favour of natural flood prevention measures such as retention areas and the re-naturalization of rivers instead of technical flood prevention. The former are not just more sustainable strategies but can also compensate for impervious surfaces, are met with more acceptability by the citizens affected and can mitigate the upstream and downstream riparian problems. Moreover, technical flood prevention strategies should be accompanied by technology assessment strategies to predict the risks. Furthermore, in order to tip the scale for sustainable flood risk management, public participation processes should happen at the beginning—not at the end—of a flood risk management process.

Initial situation and questions

The River Murr is a 51 km long tributary of the Neckar River and is considered a "flood risk river" in Baden-Württemberg due to the narrow river catchment area. It flows westwards through the Rems-Murr district and further into the district of Ludwigsburg.

Flood protection is planned by local communities and the water authorities in the Rems-Murr county district and also in the Stuttgart regional council without public participation so that civil society actors and the citizenry can not really participate. The consequences are expensive and unsustainable planning concepts based on the model of purely technical solutions of the "security strategies" of the 1950s and 1960s, without their consequences being estimated. Floodplains and nearby river areas are still being built for economic reasons, thus ensuring that the risks are passed on to the respective underlying municipalities and their citizens. Economic interests, as well as social and environmental requirements for sustainable water management and urban development, diverge widely.

This case study presents a project to promote civil society and citizenship participation in Baden-Württemberg, elaborated by the Non-Governmental Organisation Environment and Nature Conservation Association" (BUND) in 2018. The aim was to develop science-based expertise on flood protection at the River Murr, sponsored by the state's Ministry of the Federal State of Baden-Württemberg. The support of the local community was needed and carried out by the community of Steinheim/Murr in the county district of Ludwigsburg. The results of the research answer questions about the acceptance of technical flood protection by environmental associations and affected citizens and demonstrate the far higher importance of ecologically sustainable solutions through measures of water retention in the area of the river basin Murr.

The so-called "Murrtal Water Association"—consisting of the communities Murrhardt, Sulzbach/Murr, Oppenweiler and Backnang—was founded in 2008 to solve the flood problems of the River Murr. Following a 50-year flood event in 2011, the Rems-Murr County District made planning decisions in 2012 for intra-local flood protection measures for Oppenweiler in 2015 for Backnang and Sulzbach/Murr. The five flood retention basins to be built with first priority were deferred. The paradox is that in the same county, the planning on the river Rems is carried out without any internal measures and solved exclusively by flood retention basins (see Brunold 2013 and 2014).

Aspects of citizenship participation and findings of the investigation

A central component of the citizenship participation process was a standardized survey carried out by the author of this article and associated with workshops of the BUND and the NABU. The survey was carried out by a questionnaire in response to the massive internal measures of purely technical flood protection at the Murr, which now culminate in the "walling" in all four communities of the "Murrtal Water Association". The realization took place without any water retention or retention space compensation, so in the meantime, immense retention space losses of certainly more than two million cubic meters have occurred.

It is clear from this that the citizenry can often contribute greater knowledge and identification to neighbouring areas than many bureaucratic organizations, local authorities or even supervisory authorities are able to do (see Brunold 2018). The survey on flood protection at the River Murr clearly shows this. Their results will be presented below.

After looking at the actual situation through environmental data and document analyses, the dangerous risks for the citizens were first identified in workshops. The consensus was that conventional community administrative action is no longer sufficient concerning the complexity of sustainable flood risk management. For example, citizen participation should not be at the end but at the beginning of plan approval procedures, accompanied by the change from "enforcement communities" to "service and citizen communities". Innovative concepts of municipal competence requirements could contribute in such a way that administrations develop themselves into learning organisations. The added value of municipal self-government lies in its public interest orientation and its democratic participation possibilities, which can further be developed into a "cooperative democracy" in the triangle between citizenship, municipal council and administration. The ideal objective would be a symbiotic relationship between civic and civil society and public administration in order to strengthen civic participation in administrative procedures.

Ultimately, the results of the study can locally contribute to the United Nations Sustainable Development Goals. Reference points are Sustainable Development Goals No. 4 (Education), No. 11 (Sustainable Cities and Settlements) and No. 13 (Combating the Effects of Climate Change). The consequences of climate change can be demonstrated by the flood events, which are intensifying at ever shorter intervals. This requires an equitable sharing of burdens of upper and lower riparians, as can be seen by the flood risk maps of the State Institute for the Environment Baden-Württemberg. As a first result, it can be noted that the estimation of any measures taken by upper and lower riparians at the River Murr can contribute to influencing the flood situation is considered "very plausible" by 56.3% of respondents and "plausible" by 28.1% of respondents (N = 32).

One objective of the investigation is to identify planning alternatives concerning the water authorities' flood protection concepts and planning procedures. The latter shows that all community measures are preferred before the construction of flood retention basins in the exterior area, which, however, runs counter to the prohibition of deterioration in the Water Management Act in the national law. The planning of the Murrtal Water Association is based on a completely overhauled river basin survey from 2004. This was carried out—instead of using two or three-dimensional methods—with an inadequate one-dimensional non-stationary outflow calculation, which significantly distorts the results. This does not take into account the outflows of the numerous lateral waters of the Murr, so that—in view of the further increasing development of settlements—an overall balance of the conditions is not available. The "EC Directive 2007/60/EG on the Assessment and Management of Flood Risks" (EC-HWRL), which entered into force in 2007, transposed European into German law through the updating of the Water Management Act in 2010 (Federal Ministry for the Environment, 2007). The reduced quality standards are now evident in the flood hazard map 2010, which has not been updated by the State Institute for the Environment Baden-Württemberg and can be continued in a 6-year cycle. The binding deadline for updating the data by 22 December 2019 has passed.

Unfortunately, water authorities still tend to stick to the same reviewers, with almost no evaluation or certification. Compliance guidelines are also hardly taken into account. Instead, the "expertise" of the reviewers is usually taken over unaudited and confirmed. Thus, water authorities trust expensive and apparently technical feasibility rather than financially much cheaper solutions closer to nature, with the consequence that with larger order volumes, the fees of the engineering offices also increase. As a result of the externalization of environmental costs as well as subsequent care measures generated by the lack of technical impact assessments, new follow-up orders are thus always generated, and further hazard risks and costs are generated.

The survey shows a loss of public confidence in public authorities. The question of which actors have the greatest confidence in the planning of flood protection measures from the state and municipal authorities and counties (6.3%), as well as municipal associations (6.3%), performed the worst—still behind affected citizens with experience (12.5%). Environmental associations, with 50% and research institutes, with 25%, on the other hand, have the highest confidence in the planning of flood protection measures (N = 32).

It was also asked which stakeholders should be involved in the planning processes. Here the scepticism of the citizens in comparison with the authorities is again evident. Citizens' initiatives (73.9%) and research institutes (64.7%) are considered to be the main players.

Table 1.1. Key players in flood protection planning with multiple mentions (N=34).

Rank	Institutions or Actors	Frequency	Percent
1	Citizens' initiatives	25	73.5
2	Research institutes	22	64.7
3	Water associations	19	55.9
4	Water authorities	18	52.9
5	Citizens as individuals	17	50.0
6	Non-Governmental Organisations	17	50.0
7	Planning offices	10	29.4
8	Economic enterprises	2	5.9

Investigation results of the river basin area of the Murr

The land development in the river basin areas of the Murr shows the unsustainable approaches of the technical flood protection concepts. In comparison with the years 1996 and 2016, it becomes clear whether the requirements of the Water Law Baden-Württemberg and the National Water Management Act according to §5 and §6 were taken into account when draining the settlement areas. Based on data from the Federal Statistical Office of Baden-Württemberg, the land development of various municipalities in the Murr basin area are analyzed and evaluated:

Table 1: Land development in the river basin areas of the Murr from Murrhardt to Marbach.

	2016	2016	2016	2016	2016	Diff./ S+V	1996	1996	1996	1996	1996
	EW	A/ha	F+LW	S+V	W	Ha	EW	A/ha	F+LW	S+V	W
Murrhardt	13953	7115	6214	829	43	59	14346	7114	6272	770	40
Sulzbach	5194	4012	3582	387	26	28	5015	4011	3610	359	23
Oppen-weiler	4197	1980	1664	289	16	41	4081	1983	1710	248	15
Weissach i.T	7196	1413	1098	299	11	14	7028	1413	1113	285	10
Backnang	36633	3938	2722	1158	24	71	38000		2895	987	22
Aspach	8058	3545	3014	486	26	113	7960	3546	3151	373	13
Burgstetten	3653	1029	827	180	10	24	3267	1029	852	156	10
Kirchberg	3862	1321	1048	248	12	42	3657	1322	1092	206	12
Steinheim	12162	2318	1816	457	29	47	12446	2318	1838	410	30
Murr	6429	779	525	234	15	52	5432	778	572	182	15
Marbach	15682	1806	1298	448	29	20	13517	1806	1339	428	30

Legend: EW=Population, A/ha=Total area of the municipality, F=Forest areas+LW=Agricultural land, S+V=Settlement areas with traffic areas, W=Water surfaces, Diff./S+V=Difference in Settlement areas from the years 2016 to 1996.

With only small changes in the number of inhabitants from 1996 (EW 96) to 2016 (EW 16), a significantly high level of land consumption can be observed in all municipalities. This leads to significant losses in forest and agricultural land (F + LW), which is opposite to settlement and transport development (S + V).

Within an area sealing (diff./S + V) between 14 and 113 hectares, the small increase in water areas in some municipalities of about 1 to 3 hectares cannot significantly contribute to a balanced water balance or to the reduction of flood hazards there. The due diligence required in §5 and §6 of the National Water Management Act and consideration of the principles of water management is not respected by the common discharge principle of settlement drainage. In this respect, the investigation presented here, which also takes into account the flood recovery areas in the area, goes far beyond the flood hazard maps, which are mainly exclusively orientated to the course of the river.

The discharge changes in the basin area of the Murr due to the area development in the respective municipalities are shown in the lower table. In the case of a 5-year rain event (according to KOSTRA-DWD grid data) or in the case of a rain duration of 15 minutes, an average drain coefficient between traffic and settlement areas of 0.5 results in an additional drain of approximately 1 m^3/sec. for an area of 10 hectares or 100 litres/second on an area of one hectare (German Weather Service 2018).

Table 2: Changes in outflows from Murrhardt to Marbach.

City/ Community	Murr-Hardt	Sulz-bach	Oppen-weiler	Weis-sach	Back-nang	Aspach	Burg-stetten	Kirch-berg	Stein-heim	Murr	Mar-bach
Surface Sealing	+59ha	+28ha	+41ha	+14ha	+71ha	+113ha	+24ha	+42ha	+47ha	+52ha	+20ha
Discharge Coefficient	0.5	0.5	0.5	0.5	0.5	0.5	0.5	0.5	0.5	0.5	0.5
Surcharge Factor	1.1	1.1	1.1	1.1	1.1	1.1	1.1	1.1	1.1	1.1	1.1
Design Rainfall Lit./sec./ha	209.5	209.5	209.5	209.5	209.5	209.5	209.5	209.5	209.5	209.5	209.5
Rain Dura-tion/min.	15	15	15	15	15	15	15	15	15	15	15
Frequency	5	5	5	5	5	5	5	5	5	5	5
Increase Out-flow m^3/sec.	6.18	2.93	4.29	1.47	7.44	11.84	2.51	4.40	4.92	5.45	2.10
Increase Out-flow m^3/15min	5562	2640	3865	1320	6694	10653	2263	3960	4431	4902	1886

Legend: Surface sealing, difference of area development in the period of 1996 to 2016, discharge coefficient 0.5, Supplement factor for Climate 1.1, design rain following KOSTRA 2010R, set for a 5-year rain event with a duration of 15 minutes (German Weather Service 2018).

It is evident that the effects of surface sealing can be offset by the decentralized retention of surface effluents. Urgent requirements for the municipalities are needed here. Unfortunately, considerable hurdles still have to be overcome in order to implement such measures because the specifications of the National Water Management Act and the requirements in developing plans for the retention and leakage of surface drains are still too often restricted by water authorities. Furthermore, in the case of construction projects, soils are usually classified as non-leachable without differentiated investigations in order to discharge the surface water produced according to the usual principles. A balanced municipal water balance should, therefore, be prescribed in accordance with the provisions of the water laws in river basins. By creating municipal outflow and retention room balances, decentralized and regional protection concepts could be developed, which can replace the expensive and urban non-aesthetic measures—as the example of Backnang shows.

Based on drainage and retention balances, it is thus possible to comply with the requirements of the water laws. The basics are available to the municipalities through the split surface sewage fee so that local governments can account for the current data for ecologically sustainable flood protection. According to the respondents, the increased flood risks at the River Murr are, therefore, mainly due to increasing surface sealing (76.5%) and more frequent and severe rainfall events (61.8%).

Table 1.2. Assessment of the main causes of increased flood risks at the River Murr (N=34).

Rank	Cause	Frequency	Percent
1	Surface sealing	26	76,5
2	More frequent and stronger rainfall events	21	61,8
3	Settlements	15	44.1
4	Climate change	15	44,1
5	Influencing the outflow cross-section	13	38,2
6	Agricultural development	10	29,4
7	Technical flood control	9	26,5
8	Waters removal	8	23,5

This finding is closely related to the assessment of technical flood protection, which is assessed as "less sustainable" to "unsustainable," with a total of 88.8% (N = 33). The question of whether alternatives to technical flood protection are possible is answered in the affirmative by 87.5% of respondents (N = 32).

The environmental policy of the European Union (EU) aims at a high level of protection, based on the precautionary and polluter pays principle and based on the priority of addressing environmental impacts at their source. In view of the advancing European integration, it is evident that EU directive competence has an increasing influence on national laws, so a steady process of adaptation to EU law has become the rule. The problem-solving pressure within the EU multi-level system has increased and is reinforced in policy areas where public common goods can no longer be fairly distributed within national borders or solved as problems, as is the case with flood protection. This can thus be regarded as a cross-sectional task within the interface between the EU and the nation-state.

Especially in order to overcome the "prisoner's dilemma" and conflicts of goals between economics and ecology, public-interest regulations must be based on the insight that household systems are cyclical and dynamic in nature, whereas economic cycles are still perceived as linear and infinitely expandable.

When asked how important precautionary flood protection is in terms of increased flood risks, a total of 96.9% considered its relevance to be "very important" or "quite important" (N = 32). When assessing which measures are most effective for precautionary flood protection, 81.3% of the responses show that measures for restoration and land provision are preferred over technical solutions.

Table 1.3. Assessment of the effectiveness of precautionary flood protection measures (N=32)

Rank	Building measure	Frequency	Percent
1	Renaturation	16	50.0
2	Surface precaution	10	31.3
3	Technical flood control	3	9.4
4	Construction precaution	1	3.1
	Behavioral precaution	1	3.1
	Prevention of risk	1	3.1

With this finding, the effectiveness of construction measures can be evaluated for a precautionary flood protection. It has been found that 75.0% of the responses favoured measures to recover retention surfaces before the construction of technical retention solutions.

Table 1.4. Assessment of the effectiveness of precautionary flood protection measures (N=32).

Rank	Building measure	Frequency	Percent
1	Open flood surfaces	16	50.0
2	Dismantling of river banks	8	25.0
3	Constructing water-basins	6	18.8
4	Flood control polder	2	6.3

EU flood risk management and its municipal implementation

The results of the study coincide with the solutions of flood risk management, according to which public interest and locally effective regulations must be developed in accordance with EU requirements. Flood protection must therefore be seen as a Community task for the EU and for local authorities. Hereby, it is necessary to find answers to questions of justice. Thus, typical Allmende clamps and prisoner's dilemmas occur as upper- and lower-riparian problems, which pose problems for sustainable and common-good-based water management. A correction of administrative action in relation to misallocations of public subsidies and hence environmentally harmful subsidies is necessary in this field, since investments are always subject to interference with the natural water budget and to risk factors for a technology impact assessment. For example, localgovernments would have to develop an awareness of the importance of collaborations for planning and implementing sustainable flood protection measures. Cooperation is based on the confidence that each local authority is committed to implementing water laws in its respective sphere of activity with the citizenry. Last but not least, the European regulations provide for public participation. The aim of the EU Flood Directive is therefore to establish a framework for the assessment and management of flood risks to reduce flood-related consequences, in particular on human health and the environment.

After an assessment of the flood risks (Art. 4) and the determination of the areas with potentially significant flood risk (Art. 5) had to take place on 22.12.2013, the creation of flood hazard and risk maps (Art. 6) on the part of the federal states was necessary. As of 22.12.2015, the development of flood risk management plans (Art. 7,8), as well as the assessment of flood risks and flood effects on human health, the environment, cultural heritage and economic sustainability, were required. The Flood Protection Act II, which entered into force on 05.01.2018, further strengthened the flood law regulations, as well as the need to take into account neighbouring concerns and update the flood hazard maps until 22.12.2019 (German Bundestag 2017). In assessing the usefulness of flood risk management at the EU level, this was considered "very useful" by a total of 31.3% of respondents and "sensible" by 50% (N = 34).

Table 1.5. Assessment of the flood protection regime at the EU Level (N=34).

Variable	Frequency	Percent
Very useful	10	31.3
Quite useful	16	50.0
Less useful	3	9.4
Not useful	3	9.4

The EU guidelines make it clear that a wide range of aspects of river basin areas must be coordinated. Thus, risk and retention space compensation as well as the National Water Management Act's regulations and the resulting deterioration prohibitions, must be observed. However, in the federal state of Baden-Württemberg, municipalities are given opportunities by so-called "flood registers" to compensate for lost retention spaces upstream of the same river. However, there is a risk of abuse by way of exemptions, since the conditions that functionally no deterioration can occur are hardly to fulfill. Thus, the functionality of a "compensatory measure" created in 2017 in the "Upper Toswiesen" of the city of Backnang precludes the National Water Management Act prohibition on deterioration. The unlawful project "Annonaygarten" in Backnang (ZDF 2017) was an opportunity for their construction. This shows that conflicts of goals in flood protection planning are pre-programmed. 70.6% of respondents see "economic conflicts" in the foreground. 55.9% perceive "conflicts within administrative structures" as relevant, 41.2% recognize "environmental conflicts" and 35.3% fear "higher risks on the part of citizens" (N = 34).

Using the example of flood protection at the River Murr, it can be shown that ecologically sustainable and democratic action is necessary within the water authorities. Fifty percent of the respondents consider the participation of citizens in water management procedures as "very important," and 31.3% consider it "important" (N = 32). Likewise, the need for early public participation in flood protection plans of 53.1% is classified as "very high" and 28.1% as "high" (N = 32).

69.6% of the respondents agreed that subsidies for flood protection at the River Murr can harm the environment (N = 33). In this respect, environmentally harmful subsidies should be avoided there and ecologically sustainable solutions should be given preference instead (Brunold 2018). Unanimous consensus among the respondents was that any land use close to water bodies has an effect on flood events. As a counter-strategy, 63.6% of the respondents advocated the granting of subsidies for provision of land near water bodies (N = 33).

Technology assessment and its added value for flood protection on the Murr

The Murr flood protection concept shows that environmental policy and municipal decision-makers are still taking paths that run counter to the public interest orientation to the common good and the principles of ecologically sustainable development paths. Here, the technology assessment should be included in order to identify ambivalences of "technical progress" and to develop scenarios of possible technological consequences at an early stage. Taking into account expert and civic participation procedures, it is possible to derive action alternatives to overcome misconceptions in the linear and post-care-oriented consideration of flood events as well as science-based findings of natural and water cycles. Last but not least, the syndrome concept can also be applied to flood risks occurring locally, regionally and across borders, which have their causes, above all else, in rainfall-induced and increased climate change processes. Closely related to this are—as already indicated—settlement activities in the respective river basin areas, which, inter alia, impair the leaching capacity of the soils, accelerate the water outflow or tend to destroy retention areas.

The forecast scenarios of the flood hazard maps therefore reflect the consequences of technical flood protection, which, with the associated destruction of retention spaces, tend to increase the potential for danger. In view of this practice, flood hazard maps are already outdated on the day of their publication. In this respect, the highly calculated dangers in flood hazard maps can only be comprehensibly delineated to a limited extent so that future forecasts for technology assessments harbour uncertainties in knowledge and assessments, which are to be kept open, inter alia, through transparent democratic debates (Grunwald, 67 f.).

References

ARGE BAU (Hrsg.), Handlungsanleitung für den Einsatz rechtlicher und technischer Instrumente zum Hochwasserschutz in der Raumordnung, in der Bauleitplanung und bei der Zulassung von Einzelbauvorhaben, online abrufbar unter: http://www.bauen-wohnen.sachsen.de/download/Bauen_und_Wohnen/Neufassung_Handlungsanleitung_ Hochwasserschutz_2016.pdf.

Brunold, Andreas, Europäische Richtlinien und praktisches Verwaltungshandeln: das Beispiel Hochwasserschutz, in: Bundeszentrale für politische Bildung (Hrsg.), Themen und Materialien. Konzepte des Grundgesetzes—die verfassungsgemäße Ordnung der Bundesrepublik. Neun Bausteine für die schulische und außerschulische politische Bildung, Bonn 2013, S. 193-222.

Brunold, Andreas, Von der Nachsorge zur Prävention: Hochwasserschutz als öffentliche und gemeinwohlorientierte Aufgabe, in: Forum Politikunterricht. Zeitschrift der Deutschen Vereinigung für Politische Bildung—Landesverband Bayern (Hrsg.), Heft 3/2014, München 2014, S. 24-30.

Brunold, Andreas, Politische Partizipation und Bürgerbeteiligung als Kompetenzanforderung in der Landesverwaltung von Baden-Württemberg, in: Manzel, Sabine/Oberle, Monika (Hrsg.), Kompetenzorientierung. Potenziale zur Professionalisierung der Politischen Bildung. Festschrift für Georg Weißeno zum 65. Geburtstag, Springer Verlag, Wiesbaden 2018, S. 203-217.

Deutscher Bundestag: Hochwasserschutzgesetz II, Bundesgesetzblatt Jahrgang 2017, Teil I, Nr. 44, online abrufbar unter: https://www.spekter.de/doc/HochwasserschutzgesetzII-Bundesgesetzblatt.pdf.

Deutscher Wetterdienst (Hrsg.), Koordinierte Starkniederschlags-Regionalisierungs Auswertungen (KOSTRA), Datenauswertung und Datenbereitstellung durch KOSTRA-DWD- Version 2010R, Stand 2018.

Grunwald, Armin, Die hermeneutische Erweiterung der Technikfolgenabschätzung, in: TATuP—Zeitschrift für Technikfolgenabschätzung in Theorie und Praxis, Heft 2/2015, 24. Jahrgang, S. 65-69.

Landesanstalt für Umwelt, Messungen und Naturschutz Baden-Württemberg (LUBW): Umweltinformationssystem, online abrufbar unter: https://www.lubw.baden-wuerttemberg.de/umweltinformationssystem/kartendienste-der-lubw.

ZDF-Länderspiegel (2017): Hammer der Woche—Backnang baut in Überschwemmungsgebiet, online abrufbar unter: https://www.zdf.de/politik/laenderspiegel/laenderspiegel-hammer-der-woche-backnang-baut-in-ueberschwemmungsgebiet-100.html.

Chapter 9

The Take 1 Programme

Goals and Limitations depicted at the German river Murr

Valerie Lewis[1]

Abstract

Highlighting the importance of education for sustainable development (ESD) and emphasising the need for its inclusion in the formal and non-formal curriculum can seem like yet another educational challenge which educators must 'solve'. This paper focuses on providing an overview of the design and application of the Take 1 Programme of training as a mechanism for embedding ESD in learning and teaching in Irish second-level schools. It demonstrates an approach to engaging with ESD through the Sustainable Development Goals using a methodology mapped to the learning outcomes of curriculum subjects. The Take 1 Programme aims to support educators in viewing ESD as an integral part of learning and teaching, which can be encountered seamlessly throughout the learning environment.

****

As we enter the latter half of the fifteen-year timeframe of the United Nations 2030 Agenda for Sustainable Development (2015), Education for Sustainable Development (ESD) remains a focus of global challenges and has grown in relevance since the late 1980s. The role of education in achieving sustainable development is not new,

[1] Take 1 Programme Director, and Assistant Professor of Education for Sustainability, School of STEM Education, Innovation and Global Studies at the Institute of Education, Dublin City University.

and "education has long been recognised as a critical factor in addressing environmental and sustainability issues and ensuring human well-being" (GEM Report, 2016, p9), with the potential to encourage changes in knowledge, skills, and values to support a more sustainable society for everyone. The framework of 17 Sustainable Development Goals (SDGs), born from a commitment to the possibility of a new era (United Nations, 1987), provides a broad lens to support those changes. The achievement of the targets aligned with the SDGs has the potential to impact almost every aspect of wellbeing, both for humans and the planet, providing a secure and successful life for every individual.

The Take 1 Programme[2] of training is promoted as a mechanism for embedding ESD in learning and teaching in Irish second-level schools. By mapping the SDGs to the learning outcomes of subject specifications, the programme not only raises awareness for students of the SDGs but also highlights "the implications of these goals for their individual and group lives, including the responsibilities individuals and institutions have to assume to help achieve them" (UNESCO, 2020, p. 59). The Take 1 Programme demonstrates an approach that moves away from focusing on addressing specific global challenges in isolation and uses the SDGs as a framework of reference to highlight the interdependent and interconnected nature of the world in which we live.

Education for Sustainable Development

Education for Sustainable Development is not a new concept and is considered to have been formalised in its modern context via the Our Common Futures Brundtland Report in 1987, utilising a definition that continues to be referenced today—"development that meets the needs of the present world without compromising the ability of future generations to meet their own needs" (United Nations, 1987, p. 37). The critical importance of education, as aligned to achieving sustainable development, was recognised explicitly at that time. It has been a key feature of proposed additions and amendments to international ESD implementation strategies and policies in the decades since. From the subsequent UN Conference in 1992 (Rio Summit) through the Decade of Education for Sustainable Development (2005—2014),

[2] Education Matters-Ireland's Education Yearbook 2020-Take 1 Programme

the message has been consistent: "Education is critical for promoting sustainable development and improving the capacity of people to address environment and development issues" (United Nations, 1993, p. 439).

Education as a means to achieve sustainable development is very much at the centre of the current global agenda, which was unanimously adopted by all 193 United Nations member states in 2015, a move that was significant in highlighting a common pledge to action aimed at ending poverty, reducing inequality and tackling climate change by 2030. However, despite being presented in the broadest of contexts, it often defaults to or is interpreted as education as viewed in a formal sector context. Subsequently, it is the classroom teacher who is then seen as the ultimate mediator of contemporary global concerns for the young people in their charge.

ESD Policy in the Irish Context

The design and development of the Take 1 Programme has taken cognisance of the active education space in Ireland and has been driven by a desire for efficient and streamlined approaches. It has been informed by relevant whole of government policies as well as specific education policy documents, which reference education for sustainable development and the SDGs in both broad and specific contexts. The programme provides responsive and responsible opportunities that are embedded in the curriculum but impact wellbeing, which are important to the current student cohort but to future cohorts on a national and international level.

The Department of Education Strategy Statement issued for the period 2021 -2023 outlines a vision for an education system "where every child and young person feels valued and is actively supported to reach their full potential" (p. 6). It also takes cognisance of the potential of classroom learning and its impact on the wider community. The Strategy Statement acknowledges the importance of partnership, in particular cross-government objectives. It makes specific reference to its commitment to "contributing to the national effort to implement the United Nations 17 Sustainable Development Goals and sees Education for Sustainable Development (ESD) as key to achieving these" (p. 13). This is the first time a Department of Education strategy has so explicitly referenced sustainability and the SDGs. The Take 1 Programme is reflective of the aspiration to align a sense of community to a global vision by facilitating learning to permeate the whole school community and the environment beyond.

In June 2022, the Second National Strategy on Education for Sustainable Development—ESD to 2030 was launched as a collaborative effort by three government departments with education under their remit[3] and presented with an accompanying Implementation Plan to 2026 (Department of Education). The strategy spans the continuum of education in Ireland, from early learning and care to third-level education and research, and extends beyond the formal sector to engage with local communities and youth groups through lifelong learning. Language and alignment with international approaches are evident in the strategy as it sets out five key priority areas in line with the UNESCO's Framework for ESD (2019)—Advancing Policy, Transforming Learning Environments, Building Capacities of educators, Empowering and Mobilising Young People, and Accelerating Local-Level Actions.

The Second National Strategy builds on the success of the first strategy document (Department of Education, 2014) and aims to support the achievement of SDG 4.7 as well as enable the achievement of all other SDGs. It not only acknowledges the important role of SDG 4.7 but the imprint of education across all the SDGs' targets, with each of the 17 goals containing an aligned target referencing learning, educating or training, or awareness raising (GME Report, 2016). The Take 1 Programme seeks to reflect this capacity for breadth and impact across the formal education system while also supporting and encouraging learning experiences in the non-formal and informal school community. The Department of Education has reflected its support of this initiative by including the Take 1 Programme as an action of the Implementation Plan (ibid, p. 6) aligned to the strategy document under second-level targets, with a focus on continuing professional development for senior leaders in our schools.

Take 1 Programme—Alignment to National Policy

The Take 1 Programme is a professional learning programme that has been developed from a desire to support senior, middle, and teacher

[3] Department of Education, Department of Further and Higher Education, Research, Innovation and Science and, the Department of Children, Equality, Disability, Integration and Youth.

leaders in Irish second-level schools to embed ESD in learning and teaching across the junior school curriculum. The training programme was initially piloted in 2019, targeting the 252 second-level schools under the patronage of 16 Education and Training Boards (ETB) across Ireland.[4]

As part of the planning and development process for the programme, an initial period of exploration of the impact of sustainable development education within the second-level system evidenced a very diverse understanding of and engagement with the SDGs across the formal education sector. While some schools displayed in-depth knowledge, usually aligned with the work of a 'champion' teacher, the vast majority demonstrated minimal or no engagement. This observation was reflective of engagement in the wider population, with the 2018 mid-term review of Ireland's first National Strategy on ESD noting a "need to generate a greater level of awareness of ESD among all partners and citizens" (p. 44).

The challenges aligned with embedding ESD in learning and teaching in Ireland are similar to those encountered in many other jurisdictions. Despite the standard Brundtland definition, the language associated with ESD and its many elements is open to cultural and contextual interpretation. 'Sustainability', 'Sustainable Development', and 'Development' in and of themselves bring interpretations that can be wide-ranging and are more often than not associated with longevity and change. Teacher capacity is also an issue, with many acknowledging the importance of teaching ESD and global citizenship education (GCED) and their motivation to do so, "However, their drive is often overshadowed by the lack of available resources and structural support" (UNESCO and Education International, 2021, p. 6).

Significant reform in lower secondary education in Ireland occurred in 2015 with the introduction of the Framework for Junior Cycle (Department of Education and Skills). Informed by national and international research, it outlined a roadmap of engagement for young people in the first three years of second-level education and aimed to "place the student firmly at the centre of the learning process and envisages a modernised curriculum across all subjects" (Kirk, 2019, p. 192). This modernised curriculum supports multiple opportunities, on paper, to encounter ESD and related global citizenship education content across different subject areas and short courses. In

[4] Education and Training Board Schools—https://www.etbi.ie/etb-schools-ethos/

practice, teachers in second-level education settings have strong subject identities and often tend to engage in initiatives and concepts that are explicitly linked to the certainties and needs of their own subject syllabus.

The Take 1 Programme was developed on a foundation which acknowledges that the content associated with ESD and GCE can be complex and vast. Coupled with the sense of urgency generated around climate anxiety, it can be difficult to know where to begin. Training sessions capitalise on ensuring that training and information are aligned with existing and emerging education policies and practices, including the most current curriculum roadmap for junior school students. In October 2020, the Teaching Council, which is the professional standards body for the teaching profession in Ireland, published updated guidelines setting out the core elements required of Initial Teacher Education (ITE). These now include a specific requirement to incorporate GCE (including ESD) in all future teacher training programmes. This inclusion not only recognises the important role of holistic education but also highlights the timely introduction of the Take 1 Programme in upskilling teachers already in the profession.

The Take 1 Programme focuses on using the Sustainable Development Goals as the conduit for learning and teaching about the many key issues surrounding ESD. The Global Goals were chosen as an appropriate engagement point because of their universality and impact across all facets of society and as a fixed starting point which provides a common framework of reference. This connection between ESD and the SDGs was noted in the UNESCO roadmap with reference to the challenges facing the planet and an exploration of the next steps for response through education—"the criticism that ESD has received for its 'vagueness', which stems in part from its broad and inclusive perspective and the ever-changing nature of sustainability issues, can be addressed more concretely when the field is linked with the SDGs" (2020, p. 57). The roadmap considers the merger of ESD and the SDGs is not only a requirement "but a 'win-win' for both fields".

The background preparation for the Take 1 Programme training began with a comprehensive review of the syllabus content for each junior cycle subject to assess where the SDGs were referenced both implicitly and explicitly in the learning outcomes. The purpose of this review was twofold: to create a comprehensive log of potential existing ESD opportunities and to assess where additional resources might need to be created where gaps were found. The benefit of the curriculum being newly updated meant that the content was current and reflective of contemporary and emerging global citizenship concerns. Unsurprisingly, the suite of subjects that make up the Wellbeing Programme (Social, Personal and Health Education, Civic Social and Political Education, and Physical Education) contain the expected references to ESD and GCE themes. Additionally, all other subjects, without exception, contained learning outcomes which could be mapped to at least two or more of the 17 United Nations Global Goals.

This finding was pivotal, not only from a training design perspective but also from an operational and communications perspective. Guidance and interaction with participants would highlight how ESD and GCE themes were already part of prescribed learning and teaching and would not need to be presented as yet another element for inclusion. Aside from the practical advantage, this presence of ESD in the text of the compulsory national curriculum serves to move away from the construct of its narrow focus on topical issues towards a "holistic approach on learning content, pedagogy and learning outcomes" (UNESCO, 2020, p. 9). From a communications perspective, Take 1 Programme training is shared and applied using the language and terminology of the junior cycle framework document, reinforcing the capacity for embedding engagement within existing requirements and parameters. A final bonus of the inclusion of ESD across the breadth of the curriculum paves the way for whole-school engagement and collaborative practice, supporting approaches that "unfold not only through cooperative learning that occurs within the classroom but through cooperative learning between classrooms and collegial learning communities" (UNESCO, 2021, p. 83).

Take 1 Programme Training

The first training session of the cohort took place in October 2019, with participants from ETB schools self-selecting to attend an in-person day of participation. Recognising the important role of teacher leaders at both middle and senior levels, the training programme approach requires the attendance of a senior leader (Principal and/or Deputy) with teacher attendees. This rationale is informed by research which highlights the positive influence of school leadership on student learning. (Leithwood et al. 2006) (Grissom, et al., 2021). During the workshop, all participants are invited to stop and reflect at specific intervals throughout the day while also engaging in conversations about how the training might be shared on return to schools, both across subject teams and across the curriculum. Additionally, senior management participants explore embedding education for sustainable development thematically and/or at the school policy level.

Acknowledging that participants may have a range of experience, or none, about ESD, the training segments of the day provide a range of information from factual to operational, broken down broadly into the following elements.

- Background and introduction to the SDGs and their relevance to the education sector.
- How the SDGs are mapped to the learning outcomes of subjects in the Junior Cycle Curriculum.
- Exploring a planning process for embedding the SDGs in learning and teaching.
- How the Take 1 Programme aligns with Irish education policies.
- Engagement in action—planned opportunities to share and showcase learning.

While the content is comprehensive, it concentrates on providing enough information to commence the process of embedding ESD while also signposting the robust foundation and interconnected nature of the Global Goals. With an emphasis on how the SDGs are mapped to learning outcomes in all subject areas, the potential for whole school engagement is demonstrated in a flexible and bespoke manner, moving action away from being the remit of one or a few 'champion' teachers, who traditionally engaged small cohorts of students in specific activities. Following the training workshop, each participant receives all referenced and relevant support resources, including

- Subject-by-subject SDG mapping resources, which include background material and guidance to aligned ESD learning outcomes for each Junior Cycle subject;
- Background and overview notes on the SDGs from the United

Nations;

- Promotional materials on the SDGs from the United Nations;
- Resources which guide and support 'taking action" and promoting student/class/school activities;
- Promotional materials for use as part of school engagement during relevant local and national 'action' events.

The overarching intention of the Take 1 Programme training experience is to remove some of the immediate barriers to engagement and provide minimal but targeted information and resource materials. During COVID-19, the training programme was adjusted and offered online, attracting similar numbers of participants to the in-person events.

As well as highlighting how the Take 1 Programme approach showcases existing ESD learning in subject syllabuses, a section of the training demonstrates the programme's alignment to current Irish education policies and strategy standards while also mirroring international best practices. For most teachers, their engagement with policy and procedure comes to the fore during subject or whole-school inspections and as part of school self-evaluation and planning. The Take 1 Programme utilises the language of these standards where possible while also outlining how the approach adheres to the requirements of curriculum, assessment and evaluation.

In the Irish second-level setting, a set of standards against which post-primary schools can consider how well they are performing is outlined in Looking at Our School: A Quality Framework for Post-Primary Schools (LAOS) (Department of Education, 2022). Reflective of greater cohesion across the formal education sector, LAOS also supports the holistic nature of student needs and the broader aspects of learning and teaching, including those "that became all the more significant as schools responded to the needs of their students in the context of COVID-19"(p. 6). The latest iteration is referenced in the Take 1 Programme training and highlights how schools participating in the training and subsequent associated activities can demonstrate "standards of effective practice" relating to ESD as outlined in the different dimensions of the framework.

LAOS is published by the Inspectorate of the Department of Education and is considered the core framework for teachers and school leaders to be used in planning and implementing effective approaches under the dimensions of Learning and Teaching and Leadership and Management. It views career-long professional development as central to the teacher's work and firmly situates reflection and collaboration at its heart. The inspectors also use this framework to inform their work in monitoring and reporting on quality in schools. Explicit reference to sustainability has been made in LAOS 2022 and is reflective of both student capacity as well as values and policies at the management level.

In the Learning and Teaching Dimension, Domain 1: Learner outcomes (Figure 1), we see the inclusion of new text which recognises students "as key contributors to our sustainable future," and through curriculum, their understanding of ESD and global citizenship issues is developed so "they are enabled to act responsibly for a more sustainable world". (p. 27). In Domain 2: Managing the Organisation as part of the Leadership and Management dimension (Figure 2), the new text in the statements of effective practice asks that "the school values and promotes a commitment to sustainable development and implements practices and policies that embed the principles of sustainability in the daily routines of the school" (p. 38)

Standards	Statement of effective practice	Statement of highly effective practice
Students demonstrate the knowledge, skills and understanding required by the post-primary curriculum	Students are recognised as key contributors to our sustainable future. In line with the curriculum, their understanding of environmental, social and economic issues and of active citizenship, with the associated rights and responsibilities in local and wider contexts is being developed. They are enabled to act responsibly for a more sustainable world.	Students are recognised as key contributors to our sustainable future. In line with the curriculum, their understanding of environmental, social and economic issues and of active citizenship, with the associated rights and responsibilities in local and wider contexts is being developed. **Students are enabled to contribute positively, actively and compassionately towards the creation of a more sustainable and just world**

Figure 1. Statements of practice—Learning and Teaching Dimension. Domain 1: Learner Outcomes. (Looking at Our School 2022: A Quality Framework for Post-Primary Schools, p. 27)

The LAOS framework is also intended to be used by schools to inform their engagement with the school self-evaluation process (SSE). "In Ireland, SSE is designed as a process of collaborative, internal school review that is focused on school improvement" and the LAOS framework provides the criteria and standards against which schools consider how well they are performing (Inspectorate, 2022). The inclusion of an explicit reference to environmental and social issues aligned with active citizenship is pivotal in acknowledging the role of ESD in contemporary Irish education.

Standards	Statement of effective practice	Statement of highly effective practice
Manage challenging and complex situations in a manner that demonstrates equality, fairness and justice	The school values and promotes a commitment to sustainable development and implements practices and policies that embed the principles of sustainability in the daily routines of the school	The school **prioritises and promotes, as a core value**, a commitment to sustainable development, **the school community collectively** develops and implements practices and policies to embed principles of sustainability in the school's daily routines.

Figure 2. Statements of practice—Leadership and Management Dimension. Domain 3: Managing the Organisation. (Looking at Our School 2022: A Quality Framework for Post-Primary Schools, p. 38)

In addition to referencing these policies and frameworks throughout the Take 1 Programme, the training workshop demonstrates practical examples of how learning and teaching about ESD can be seamlessly embedded in current teaching approaches and in school planning, assessment and evaluation practices. The foundational concept of the Take 1 Programme philosophy is the ability to engage with ESD in a built-in manner. "It is interdisciplinary and transdisciplinary, meaning that no discipline can claim ESD as its own, but all disciplines can respond and contribute to ESD individually and/or collectively" (UNESCO, 2018, p. 35).

The Take 1 Programme In Action

The learning theory most relevant to ESD is transformative learning theory (Clark, 1993), which concentrates on transforming individual perceptions through three dimensions: psychological, convictional and behavioural. The behavioural dimension—how individuals change the way they respond to their physical environment—is relevant to ESD because it provides insights into how to restructure the relationship between humans and their environment. The Take 1 Programme training workshop provides participants with an introduction to ESD and the SDGs, resource materials for individual subjects that map the SDGs to learning outcomes, and the relevant policy alignment rationale for planning and evaluation. However, the workshop is considered to be the start of the ESD journey at the individual and school level. The training and information offered at each session provide a toolkit of knowledge which builds the capacity of educators to facilitate students' transition to sustainable ways of life. "In order to guide and empower learners, educators themselves need to be empowered and equipped with the knowledge, skills, values and behaviours that are required for this transition" (UNESCO, 2020, p.30).

Following Take 1 Programme training, participants are encouraged to share their learning, both with their teacher colleagues and students and to seek an opportunity to put that learning into action. While learning and activities can take place in individual classrooms, schools are encouraged to engage in a collaborative approach, both through sharing initial resource materials and by working collectively through a whole-school approach. One of the many strengths of the Take 1 Programme is that it is highly inclusive of all learners. As engagement is mapped to learning outcomes, no subject-specific modifications are needed, and no adjustment is made for gifted learners or those with special needs, nor senior or junior students. Learning and teaching about ESD and the SDGs and their relevance to contemporary subject content occurs as and when relevant subject content alignment appears. Individual teaching plans do not need to be adjusted or amended as teachers reference sustainability in an inbuilt and streamlined manner.

Figures 3 and 4 provide a sample of the SDG subject mapping resource associated with two junior cycle subjects, Home Economics and Business Studies. The Strands and Learning Outcomes are set by the National Council for Curriculum and Assessment (NCCA) and apply to all students studying these subjects. The learning outcomes listed are those identified by the author as having an implicit or explicit relationship with the SDGs, represented by the icons at the top of each figure. During the training workshop, it is noted that these are just a selection of learning outcomes as a starting point for engagement, and it is likely that the subject teacher, who is familiar with the details of each syllabus, will identify even more examples.

The benefits of this approach are many: demonstrating ESD concepts as an embedded construct, interwoven throughout the learning; exposing all students in the class group to ESD learning; and demonstrating that ESD is the remit of all subjects, moving away from a siloed and subject-specific approach. Additionally, this methodology responds to students' concerns and entitlement to learning about and for sustainability while also allowing each school to move forward in a way that acknowledges its own needs and contexts. It has been evidenced that incorporating ESD into existing curricula helps improve students' quantitative reasoning and scientific understanding and enables them to apply their critical thinking abilities and problem-solving skills to the multifaceted and complex issues of global sustainability (United Nations, 2015).

Home Economics

Strand 1	Strand 2	Strand 3
Food, Health & Culinary Skills	**Responsible Family Living**	**Textiles and Crafts**
Learning Outcomes	Learning Outcomes	Learning Outcomes
1.15 investigate the impact of their food choices from an ecological and ethical perspective *1.16 apply sustainable practices to the selection and management of food and material resources*	*2.7 identify how individuals, families and households can contribute to sustainable and responsible living* *2.8 describe sustainable everyday practices in the home to include energy efficiency, waste management and water conservation* *2.9 explore the influence of technology on the management of personal, family and household resources*	*3.6 demonstrate ways in which clothing and/or textile household items can be repaired, re-used, re-purposed, recycled and upcycled* *3.7 evaluate textile care procedures used in the home from an environmental perspective* *3.8 discuss the influences of trends and choices on textile and clothing, including ethical and ecological considerations*

Figure 3. SDG Mapping to Learning Outcomes in Home Economics Junior Cycle Curriculum Take 1 Programme Training Resources

ESD learning and teaching at the classroom level can be further enhanced by collaboration through whole school engagement. While it is emphasised that Take 1 Programme engagement is not time-bound, two national showcasing opportunities are coordinated, which offer schools, teachers and students an opportunity to share their ESD activities. These showcasing events usually follow a training workshop and are presented as a platform to encourage participants to work towards putting their own learning into practice. Where schools have little or no previous engagement with ESD, these events offer a chance to launch or commence actions. Where schools have a more embedded approach, these occasions offer an opportunity to re-energise and celebrate existing actions and activities.

Business Studies	8 Decent Work and Economic Growth · 9 Industry, Innovation and Infrastructure · 10 Reduced Inequalities · 11 Sustainable Cities and Communities · 12 Responsible Consumption and Production	
Strand 1	Strand 2	Strand 3
Personal Finance	**Enterprise**	**Our Economy**
Learning Outcomes	Learning Outcomes	Learning Outcomes
1.2 Identify and classify sources of income and expenditure, compare options available to best manage financial resources, evaluating the risks associated with each option and making informed and responsible judgements *1.9 Debate the ethical and sustainability issues that arise from their consumption of goods and services and evaluate how they can contribute to sustainable development through consumer behaviour*	*2.2 Describe the skills and characteristics of being enterprising and appreciate the role of an entrepreneur in an organisation, in society and to the economy* *2.5 Investigate the positive and negative impacts on a community of an organisation from an economic, social and environmental perspective*	*3.1 Explain how scarcity of economic resources results in individuals having to make choices; predict possible consequences of these choices* *3.6 Explain how economic growth can impact positively and negatively on society and the environment and justify the promotion of sustainable development* *3.11 Evaluate the benefits and costs of a government economic policy and assess who enjoys the benefits and who bears the costs*

Figure 4. SDG Mapping to Learning Outcomes in Business Studies Curriculum. Take 1 Programme Training Resources

Take 1 Week has taken place during the second half of November since the inception of the Programme in 2019. While a specific week is announced each year, schools are free to engage prior to or after the allocated time frame. This flexibility acknowledges that schools are busy environments working towards end-of-year activities and assessments and prioritises participation even if it may not align with the assigned calendar week.

During Take 1 Week, teachers are asked to Teach 1 Lesson, about 1 SDG, to 1 Class Group at any point in their regular timetable of classes. (Figure 5) This is a minimum engagement, and teachers are free to do more should the opportunity arise. They are encouraged to work within their planned scheme of work, with no requirement to pause or amend regular activities or content. This approach emphasises the permeated nature of ESD in subject content and curriculum. Additionally, if whole-school participation can be encouraged, students are potentially exposed to numerous experiences of learning about ESD throughout the school environment as they move from subject to subject and classroom to classroom. Not only does this approach demonstrate the broad interdependent nature of sustainability and global citizenship issues to students, but it also highlights their imprint on all subject matter and establishes a collaborative example of communities of practice.

Figure 5. Promotional poster for Take 1 Week 2022

Events and activities are shared on Twitter/X's social media platform. This format was chosen to provide an immediate and uncomplicated space to share what is happening throughout the various school environments. The limited maximum character length for posted messages and the capacity to include images and/or short videos means that participants do not have to prepare elaborate reports on engagement but can provide an immediate and vibrant snapshot. Using an aligned hashtag (e.g. #etb_sdgs and #take1week2022) ensures that posts can be collectively found and schools can view, like, and share examples of learning and teaching about ESD. This space has also become a useful reference for schools that are new to the process, providing samples of the breadth and range of activities from around the country.

From a monitoring and evaluation perspective, sharing in this way provides the author with an avenue to collect statistical and narrative data, which can be used to follow progress as well as update and improve the programme. This is in addition to more formal evaluation through gathered data following each training session. A social media presence also provides Take 1 with the capacity to reach beyond the sector, demonstrating engagement and partnerships with external agencies, particularly at the community level. Additionally, it provides accessible public updates on how our schools are engaging with education for sustainable development and learning from each other.

The second engagement episode occurs annually in early May and commenced in May 2021 during the COVID-19 pandemic. Second-level schools in Ireland had been closed since January of that year, and Take 1 Programme training had moved online. May Day for Sustainability was introduced as a focused single day of celebration of engagement and concentrated on following up from Take 1 Week as well as providing an opportunity to present some outdoor ESD activities. These events are again shared on social media with many schools concentrating on good health and wellbeing and environmental issues such as biodiversity, and water conservation. (#mayday2022)

While the Take 1 Programme training sessions provide an opportunity for professional development aligned to ESD, the immersive and inclusive methodology acknowledges the holistic nature of education. There is no directed pedagogy promoted throughout the programme, but instead, teachers are offered an opportunity to situate ESD in their own context, both at the classroom and school level. A narrow focus on topical issues of concern may create the appearance of a responsive and current education environment, but it is less likely to bring about a transformational behaviour shift. A 2020 UNESCO study which reviewed country policy documents showed that "ESD is mostly associated with the teaching of scientific knowledge on environment" (p. 9), with Chung and Park noting that "sustainability is not likely if focused only on environmental issues such as climate change and resource depletion" (2016, p. 20) If we are to promote and value the interdependent nature of our impact and progress, we must continue to support a broad and inclusive attention to all ESD issues.

Exposing every student in every classroom to ESD in every subject not only presents sustainability and global citizenship issues as inclusive and interdependent but also supports an alternative avenue for expressing student voice and participation. While many of our student cohorts have engaged in climate strikes as a demonstration of their desire for action, many more do not have the capacity or the desire to engage in this way. This is not to negate this approach, but rather to acknowledge that student 'voice' can be expressed in many formats and many guises. If we truly aspire to be inclusive, then we must present and acknowledge inclusion in its broadest possible interpretation. We must seek out and listen to the lesser heard voices, and acknowledge that poverty and fragility can exist in local as well as global contexts. We must not lose sight of the fact that environmental issues and climate concerns do not exist in isolation and that they are connected to and addressed by attending to other global goals such as good health and wellbeing and gender inequalities.

Conclusion

Embedding ESD in learning and teaching has proved challenging, and research undertaken over the last two decades has proposed many collective and individual explanations. Competition for time in an already overcrowded curriculum tends to relegate ESD to an elective position (Ballantyne, 1995), and despite its visibility and presence in the new Junior Cycle Framework, manifesting it as more than a light touch can seem elusive. The Covid pandemic has compounded these curriculum issues, where prescribed content was condensed and prioritised in the adapted online space, and teachers are still adjusting to the return to in-person education. The pandemic had an additional impact on many students, highlighting the impact of loss of close contact with teachers and friends, as well as "the wide variety of educational, social and health-related services which schools offer" (OECD, 2022, p. 19)

Ironically, it is because of these challenges that learning for ESD has never been more important, and the Take 1 Programme provides a methodology to do this in an embedded way. We must move away from the notion that ESD and teaching about the SDGs is only the remit of the teacher educator with expertise in sustainability. Learning and teaching about and for sustainable development across a range of disciplines communicates that sustainable development is essentially a collective endeavour for schools and wider society, whether in formal, non-formal or informal learning spaces. We must also apply that collective endeavour across the spectrum of global goals. While SDG 13 (Climate Action), including environmental education and biodiversity, seem to be the sustainable priorities of the moment, it would be more productive to operate from a broader perspective.

The 2030 Agenda for Sustainable Development contains a pledge that "no one will be left behind," this mission of inclusivity should remain the singular focus for success. Should we promote a champion for this aspiration, it might be worth a closer look at the text aligned to SDG 17, emphasising policy, coherence, planning, and a desire to revitalize the Global Partnership for Sustainable Development (United Nations, 2015). The Take 1 Programme echoes many of these aspirations, reflecting national education policies, coherence with the prescribed curriculum, and an alignment to recommended planning and assessment frameworks. However, its capacity for collaborative engagement and transformational learning is its highest aspiration, demonstrating how ESD connects us, from classroom to classroom, throughout the school, and into the wider community, in a hopeful demonstration of a more sustainable future for all.

References

Ballantyne, R. (1995). Environmental Teacher Education: Constraints, Approaches and Course Design. *International Journal of Environmental Education and Information*, *14*(2): 115-128.

Clark, M.C. (1993). *Transformational learning. New Directions for Adult and Continuing Education*, 1993 (57): 47-56. Available at: https://doi.org/10.1002/ace.36719935707

Chung, B and Park, I (2016). A Review of the Differences between ESD and GCED in SDGs: Focusing on the Concepts of Global Citizenship Education. Seoul National University, Republic of Korea CICE Hiroshima University, *Journal of International Cooperation in Education*, *Vol.18* No.2 (2016) pp.17 - 35

Department of Education (2014). *1st National Strategy on Education for Sustainable Development in Ireland, 2014-2020.* Dublin: Department of Education. Available at: https://www.gov.ie/pdf/?file=https://assets.gov.ie/70926/7e8f9ccf9069441e960d7f32a38dc7b9.pdf#page=null

Department of Education and Skills (2015). T*he Framework for Junior Cycle.* Dublin: Department of Education and Skills. Available at: https://ncca.ie/media/3249/framework-for-junior-cycle-2015-en.pdf

Department of Education (2018). N*ational Strategy on Education for Sustainable Development in Ireland—Report of Interim Review and Action Plan for Q4 2018-Q4 2020.* Available at: https://www.gov.ie/pdf/?file=https://assets.gov.ie/70925/6dbd95ae3caa40b6a38a118ea669d5bc.pdf#page=null

Department of Education (2021). *Strategy Statement 2021 -2023.* Dublin: Department of Education. Available at: https://www.gov.ie/en/publication/56137-department-of-education-statement of-strategy-2021-2023/

Department of Education (2022a). *2nd National Strategy on Education for Sustainable Development—ESD to 2030.* Dublin: Government of Ireland. Available at: https://www.gov.ie/pdf/?file=https://assets.gov.ie/228330/c69895a6-88f0-4132-b6d1-9085a9c31996.pdf#page=null

Department of Education (2022b). *ESD to 2030: Implementation Plan 2022–2026.* Dublin: Government of Ireland. Available at: https://www.gov.ie/pdf/?file=https://assets.gov.ie/228358/86ea1f5f-ff88-45f0-8ee5-270997007db1.pdf#page=null

Department of Education (2022c). *Looking at Our School 2022: A Quality Framework for Post-Primary Schools.* Available at: https://www.gov.ie/en/publication/b1bb3-

Grissom, J.A., Egalite, A., and Lindsay, C., (2021). *How Principals Affect Students and Schools: A Systematic Synthesis of Two Decades of Research.* New York: The Wallace Foundation. Available at: http://www.wallacefoundation.org/principalsynthesis

Inspectorate (2022). *School Self-Evaluation: Next Steps September 2022—June 2026. Dublin: Department of Education.* Available at: https://www.gov.ie/en/circular/e0383-school-self-evaluation-next-steps-september-2022-june-2026/

Kirk, P., (2019). *Junior Cycle Reform: The Reality of Classroom-Based Assessment CBA),* in Ireland's Yearbook of Education 2019—2020. Ireland: Education Matters pp. 192-196

Leithwood, K., Day, C., Sammons., P., Harris, A. and Hopkins, D. (2006) *Seven strong claims about successful school leadership.* Nottingham: National College for School Leadership.

OECD—*Organisation for Economic Co-operation and Development (2022).* The Impact of Covid-19 on Education—Insights from Education at a Glance. Paris: Organisation for Economic Co-operation and Development.

Take 1 Programme: Embedding Education for Sustainable Development in Learning and Teaching (n.d.). Available at: https://www.take1programme.com/

The Teaching Council (2020) Céim: Standards for Initial Teacher Education. Maynooth, Ireland: The Teaching Council. Available at: https://www.teachingcouncil.ie/en/publications/ite-professional-accreditation/ceim-standards-for-initial-teacher-education.pdf

United Nations (1987). *Report of the World Commission on Environment and Development* : note / by the Secretary-General. New York: United Nations.

United Nations (1993). *Report of the United Nations Conference on Environment and Development. Volume I: Resolutions Adopted by the Conference.* New York: United Nations.

United Nations (2015). *Transforming Our World: The 2030 Agenda for Sustainable Development.* United Nations: New York Available at: https://unesdoc.unesco.org/ark:/48223/pf0000261445.

United Nations System: Chief Executives Board for Coordination (2017) *Leaving No One Behind: Equality and Non-Discrimination at the Heart of Sustainable Development.* A Shared United Nations System Framework for Action. New York: United Nations.

UNESCO and Education International (2021). *Teachers have their say: Motivation, skills and opportunities to teach education for sustainable development and global citizenship.* United Nations Educational, Scientific and Cultural Organization: Paris and Education International: Belgium.

UNESCO (2018). Leicht, A., Heiss, J. and Won Jung, B. (eds) *Issues and Trends in Education for Sustainable Development.* Paris: United Nations Educational, Scientific and Cultural Organization.

UNESCO (2020) *Education for Sustainable Development: A roadmap.* Paris: United Nations Educational, Scientific and Cultural Organization. Available at: https://unesdoc.unesco.org/ark:/48223/pf0000374802

UNESCO (2021) *Reimagining our futures together—A new social contract for education.* Report from the International Commission on the futures of education. Paris: United Nations Educational, Scientific and Cultural Organization.

UNESCO (2022). *What you need to know about education for sustainable development* (Updated, June 9th) Available at: https://www.unesco.org/en/education/sustainable-development/need-know

Chapter 10

Development and Trial of Climate Change Education Program Featuring "Carbon-neutral Simulator" for School

Hideaki Kurishima[1], Rumi Yatagawa[2], and Hidefumi Kurasaka[3]

Abstract

To achieve carbon-neutral regions, younger generations must take responsibility at the regional level and understand the pathway toward decarbonization while considering appropriate countermeasures. However, decarbonization is a complex concept for children, making it challenging for them to discern which countermeasures and technologies are effective. Therefore, it is necessary to design climate change education (CCE) programs for children. In this study, we developed a school version of a "carbon-neutral simulator (CNS)" application that can explore carbon-neutral strategies at the regional level in 2050 and implemented a CCE program using the application on a trial basis. When students were surveyed after attending a class using the CNS, the majority reported being able to operate the CNS easily and understand the decarbonization process. Furthermore, many students expressed their ability to operate the application independently and enjoyed the learning process. Through this education program and application, students were able to grasp climate change and decarbonization as "my problem".

Keywords: climate change education (CCE), decarbonized society, simulator, regional scale, junior high school

[1] Professor, School of Architecture, Shibaura Institute of Technology, Japan
[2] Professor, College of Engineering, Shibaura Institute of Technology, Japan
[3] Professor, Graduate School of Social Science, Chiba University, Japan

Introduction

An increasing body of scientific evidence highlights climate change and its effects as the world strives to rapidly decarbonize. In 2015, the Paris Agreement established its goal "to limit [anthropogenic] global warming to well below 2, preferably to 1.5 degrees Celsius" (UN FCCC, 2015). Subsequently, the IPCC published its *Special Report on Global Warming of 1.5°C*, which stated that "the impacts of 1.5 °C and 2 °C are completely different" and that "in order to keep the temperature rise within 1.5 °C, we must reduce greenhouse gas emissions to net zero from 2040 to 2055" (IPCC, 2018). In response, countries accelerated their decarbonization targets by revising their Nationally Determined Contributions (NDCs). In October 2020, Japan announced its national goal to attain a domestic carbon-neutral society by 2050 (Prime Minister's Office of Japan, 2020). Nonetheless, the actions of various nations have remained insufficient.

Initiatives at the regional level are indispensable for realizing a decarbonized society. As of 2022, approximately 800 regional governments in Japan have committed to net-zero carbon emissions by 2050 (MOE, 2022). However, few municipalities have enacted specific climate change mitigation strategies to reduce GHG emissions in their regions. This is because many regional communities and businesses do not attach much importance to realizing a decarbonized society. A survey conducted by the Japanese government in 2020 revealed that only 19.1% of respondents were aware of the details of the Paris Agreement, and 33.2% understood the concept of a decarbonized society (MOE, 2021). Globally, there is a rising concern over the problems associated with climate change, but some people do not see it as a severe threat to themselves. Realizing carbon-neutral regions requires raising public consciousness of a decarbonized society. It is essential for the younger generations responsible for future regional societies to understand the path toward decarbonization and consider effective countermeasures. However, decarbonization is not a simple concept for children, making it challenging to grasp effective countermeasures and technologies. Therefore, it is necessary to design climate change education programs for children.

Interest in climate change education (CCE) has increased recently, leading to numerous implementation and research efforts. Monroe et al. (2017) conducted a literature review of CCE. They identified the following two elements for its effectiveness: (1) a focus on personally relevant and meaningful information and (2) the use of active and engaging methods. Regarding (1), the place-based approach can help learners comprehend the global and regional climate change context and recognize climate change as "their issue" (Hallar et al., 2011). Quantitative examples demonstrate the amounts of greenhouse gas emissions produced by various activities (Cordero et al., 2008). Existing CCE programs have merely conveyed scientific facts on climate change or attempted to link them to changes in individual behavior. Concerning (2), the effectiveness of learner-centered and inquiry-based methods (Brumann et al., 2022), group deliberative discussions (Klosterman and Sadler, 2010), and role-playing and simulation (Sterman et al., 2014) has been demonstrated. However, we believe there has been insufficient implementation and research on inquiry-based programs incorporating simulations and discussions centered around specific regional policies to achieve future carbon neutrality.

Based on the above considerations, this study developed and implemented a CCE program in which students explored methods for achieving future carbon neutrality in their local areas using simulations to raise awareness of a decarbonized society among children (middle and high school students). In Japan, the Ministry of Education, Culture, Sports, Science and Technology (MEXT) is currently promoting the "GIGA School Program," intended to provide "one device for one student" is being promoted by (MEXT, 2020). This program was hastened by the rising need for at-home learning due to the COVID-19 pandemic. Therefore, we developed a school version of a carbon-neutral simulator (CNS) application software that can explore regional carbon-neutral for 2050 and created an education program using the application. Students envision themselves as city mayors using this program and explore ways to promote regional decarbonization by 2050. Through simulations, they can assess the greenhouse gas reductions achievable by implementing various policies and gain firsthand experience of their effects. Thus, they can obtain a deeper understanding of decarbonization than by simply attending lectures.

Galeote et al. (2021) reviewed the use of games and gamification in CCE and discussed their effectiveness. "Climate Action Simulation" is an existing CCE program intended to create a decarbonized society. Learners' role-play is based on computer simulation information

(Rooney-Varga, et al., 2020). Its implementation enhanced learners' knowledge of decarbonization measures and heightened their awareness of contributions to solving climate change problems. However, this program considers carbon neutrality at the global level, a scale that children may find difficult to relate to. Conversely, our program considers carbon neutrality at the regional level, aligning with learners' contexts.

Climate Change Education Program Featuring "Carbon-neutral Simulator" for School

The purpose of this CCE program is twofold: 1) to conduct an independent study of the mechanisms and effects of climate change as a global and regional issue, along with the technology and systems to address it using regional resources, and 2) to empower students to explore ways of achieving decarbonization in their respective regions by 2050. This is expected to help students acquire knowledge about climate change, regional resources, energy, and decarbonization and foster motivation and interest in realizing decarbonization in the future.

The program we have developed consists of the following four-hour block. The first hour is a lecture on the "Climate change mechanism and its effects." Here, students first learn about the present state of climate change and its mechanisms based on data from IPCC reports (IPCC, 2021). Next, data from IPCC reports (IPCC, 2021, 2022), Japan Meteorological Agency (JMA) reports (JMA, 2020), and the Climate Change Adaptation Information Platform (A-PLAT) of the Ministry of the Environment (MOE, n.d., a) were used to learn about the effects of climate change on the world, Japan, and the regions where students live. Specifically, they learned about the impact of disasters on human health, ecosystems, agriculture, livestock, and fisheries. Furthermore, they investigated whether climate change affected their regions.

The second hour is a lecture on "Mitigation: Ways to reduce greenhouse gas emissions." Here, students first learn about movements toward decarbonization at the global and regional levels in Japan, such as the Paris Agreement, Japan's NDC, and the zero-carbon declarations by local governments in Japan. Next, they learn about energy savings, renewable energy technology and systems, and carbon sinks as a means of reducing greenhouse gases. Specifically, they learn about the "Net-Zero Energy House (ZEH)" and "Net-Zero Energy Building (ZEB)" that eliminate energy input from the outside of buildings; reduce automobile use through the compact city concept; vehicle electrification; renewable energy such as solar, wind, small hydropower, biomass, and geothermal power generation; and carbon dioxide absorption by forests.

The third hour is the "2050 Carbon-neutral experience" using CNS. Details of the CNS are explained in the following section. First, students experience methods of achieving decarbonization in the regions where they live using the CNS on their own devices. The CNS can simulate the various technologies and systems connected to carbon dioxide reduction learned during the second hour. CNS outputs determine whether regional carbon neutrality can be achieved, as well as the costs involved. Thereafter, the results were shared with other students, and regional carbon neutrality was discussed within the groups.

The fourth hour is "Exploring and presenting methods of simultaneously solving decarbonization and regional issues." In addition to decarbonization, other issues facing regions are declining regional industry, population decline, and aging. The groups have discussed methods to solve these issues and decarbonization simultaneously. During this time, the results of the CNS used in the third hour were also referenced. Finally, specific policies are presented to solve decarbonization and regional issues simultaneously.

Thus, students are first taught the fundamental science of climate change and decarbonization at global and regional levels. Their understanding of regional decarbonization was further deepened through the experience of the CNS. Furthermore, the students discussed and presented simultaneous solutions to regional issues and decarbonization. Put simply, this program leads students from knowledge to experience, experience to thought, and thought to expression. This is expected to establish literacy and competencies related to climate change and decarbonization.

Carbon-neutral Simulator (CNS) for School

Structure of CNS

The CNS is a tool that can easily simulate measures for each municipality to achieve carbon neutrality in the household, transportation, and business sectors by 2050. The structure of the simulator is illustrated in Figure 1.

(1) The residential and commercial floor areas and the number of private and commercial vehicles in 2050 were estimated based on the *Regional Population Projections for Japan* of the National Institute of Population and Social Security Research (2018). This estimation result was calculated without changing the present floor area per population or the number of vehicles. These data were then used to estimate the business-as-usual (BAU) carbon footprint for 2050. Furthermore, the BAU scenario used current performance data for energy consumption per building floor area, fossil fuel consumption per vehicle, and carbon dioxide emissions per kWh of electricity.

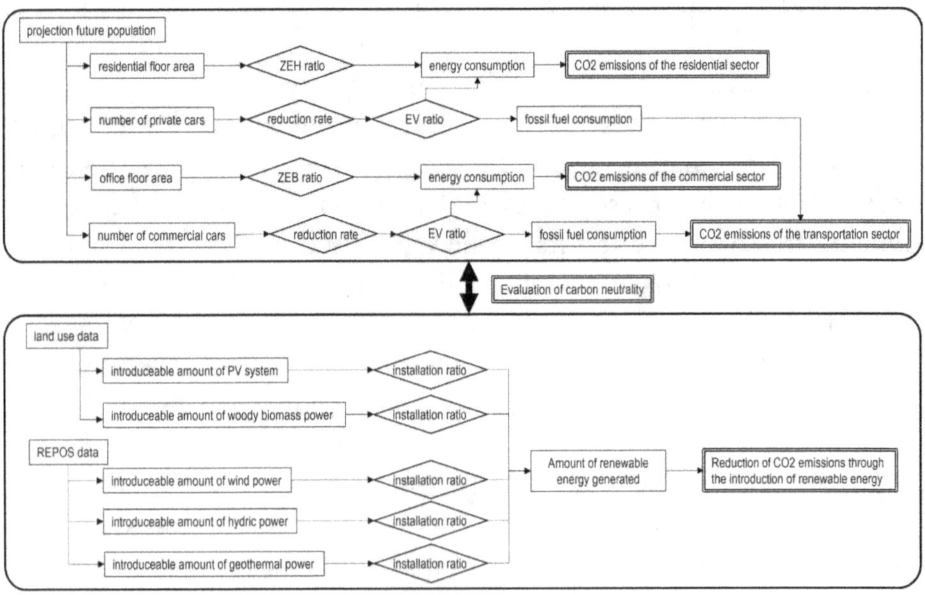

Figure 1. Structure of Carbon-Neutral Simulator

(2) The area of buildings to be reconstructed within a municipality by 2050 was estimated from the construction age of existing buildings. Subsequently, the energy consumption of the residential and commercial sectors was calculated by considering the percentage of buildings that will remain standing between now and 2050, as well as the buildings replaced by 2050 that will become "ZEHs" or "ZEBs." Further-

more, the energy consumption of the ZEH and ZEB was calculated as zero.

(3) By inputting the reduced number of vehicles estimated in (1) and the rate of vehicle electrification, both electric power consumption and the amount of carbon dioxide derived from fossil fuels in the transportation sector by 2050 were estimated.

(4) The amount of solar power that can be introduced is estimated from developed land areas, such as parking lots, vacant land, abandoned farmland, and farmland in a municipality. The amount of woody biomass power generation that could be introduced was estimated based on the forest coverage of a municipality. The amount of onshore wind, hydroelectric, and geothermal power that can be introduced into a municipality is derived from the Renewable Energy Potential System (REPOS) data from the MOE (MOE, n.d., b). Additionally, renewable energy unique to a region (e.g., agricultural biomass) is sometimes added as an option. The amount of carbon dioxide emission reduction compared to thermoelectric power generation by fossil fuels was then estimated by inputting the percentage of each form of renewable energy introduced.

(5) The achievement of local carbon neutrality is determined based on the results in (2), (3), and (4). Furthermore, total investment, energy cost reduction, and renewable energy sales were calculated simultaneously.

Development of CNS for School

The CNS was designed as a Microsoft Excel application. The CNS was originally developed to allow municipal policymakers to formulate regional decarbonization strategies. Figure 2 illustrates the initial simulator input and output screens. The inputs included the projected implementation levels of energy conservation and renewable energy measures for the 2020s, 2030s, and 2040s. Based on these inputs, the simulator estimated the resulting changes in carbon dioxide emissions.

In August 2020, we conducted an event and tested the prototype with 51 middle and high school students. However, most of the responses from both students and teachers were negative, such as "Difficult to operate," "It is difficult to know what I am supposed to input," and "It is difficult to understand the content." Therefore, it is necessary to develop a school version of the CNS. We developed a school version

	Energy Saving/Renewable Energy Investment Options	Investment unit	Investment per unit (million yen)	CO2 rediction (t-CO2)	Unit investment in the 2020s	Investment in the 2020s (million yen)	Maintenance cost, etc (million yen)	Useful life (year)	Power sales	Fuel cost reduction	Required quantity	Unit	Residual quantity	Introduceable quantity		
A	Make new houses ZEH	100 hoses	220	309	6	1,320	0	50		○	600	hoses/10years	152	752	hoses/10years	Estimated number of newly built houses
B	Energy-saving renovation of existing houses	100 hoses	60	59	10	600	0	50		○	1000	hoses/10years	11,094	12094	hoses/10years	Number of non-energy-saving houses
C	Installation of PV(4kW) in existing houses	100 hoses	160	227	3	480	0	20		○	300	hoses/10years	3,102	3402	hoses/10years	Number of homes without PV installed
D	Installation of solar water heater in existing houses	100 hoses	40	45	2	80	0	26		○	200	hoses/10years	3,370	3570	hoses/10years	Number of homes without solar water heater installed
E	Solar power plant(100KW) + Storage(200kWh)	1 unit	34.2	57	3	103	20	20	○		1.2	ha	171	173	ha	30% of abandoned farmland area, 50% of vacant land/parking lot area
F	Solar power plant(100KW) + Storage(200kWh) *Foreign investment	1 unit	3.42	57	3	10	0	20	○							
G	Waste cooking oil biodiesel plant (13,800liter)	1 unit	17	20	1	17	10	20		○	13800	liter/year	8,200	22000	liter/year	Amount of waste cooking oil
H	Wind power plant (2,000kW)	1 unit	564	2,024	5	2,820	1,050	20	○		10000	kW	401,400	419400	kW	resource potential
I	Wind power plant (2,000kW) *Foreign investment	1 unit	56.4	2,024	4	228	0	20	○		8000	kW			kW	
J	Wood chip boiler (240kW)	1 unit	277	680	1	277	59	20	○	○	600	tons/year	9,400	10000	tons/year	Amount of wood chips
K	Sugar cane bagasse power generation (1,000kW)	1 unit	200	2,834	1	200	50	20	○	○	32336	tons/year	0	32336	tons/year	Amount of sugar cane bagasse

Total investment in the 2020s	7322	million yen
Amount investable in the 2020s	7661	million yen
The balance in the 2020s	339	million yen

Please decide the unit investment in the 2020s so that "the balance" and "residual quantity" will not be negative.

			occurrence
Variable factor in the 2020s	Event 1	Movement against foreign investment	○
	Event 2	Cannot secure forestry workers	
	Event 3	Strong Typhoon Annual	○
	Event 4	Soaring fossil fuel prices	

	2020	2030	2040	2050
CO2 emission (t-CO2/year)	139,624	118,225	97,659	76,761
CO2 reduction amount (t-CO2/year)		25,299	65,035	78,166
CO2 reduction ratio		21.4%	66.6%	101.8%

	2020	2030	2040	2050
Photovoltaic (TJ/year)	170	219	312	575
Wind power (TJ/year)	13	387	1,010	623
Biomass power (TJ/year)	0	72	142	282
Solar heat utilization (TJ/year)	33	103	456	1,447
Energy demand (TJ/year)	2437	2067	1709	1345
Energy self-sufficiency rate	8.9%	37.8%	112.4%	217.5%

Figure 2. CNS Prototype Input (Top) and Output Screens (Bottom)

of the CNS while referencing the opinions of students and teachers. Figure 3 shows the input/output screen of the developed CNS school version.

First, the input and output screens were integrated into the school version of the CNS. Furthermore, the input and output items were reduced compared to the version for policymakers. Additionally, only carbon dioxide emissions and reductions in 2050 were output in the form of a graph. These improvements made the application easier to use and understand. The school version of the CNS can be downloaded from our project website. Furthermore, the CNS can be used not only in Microsoft Excel but also in Google Spreadsheets.

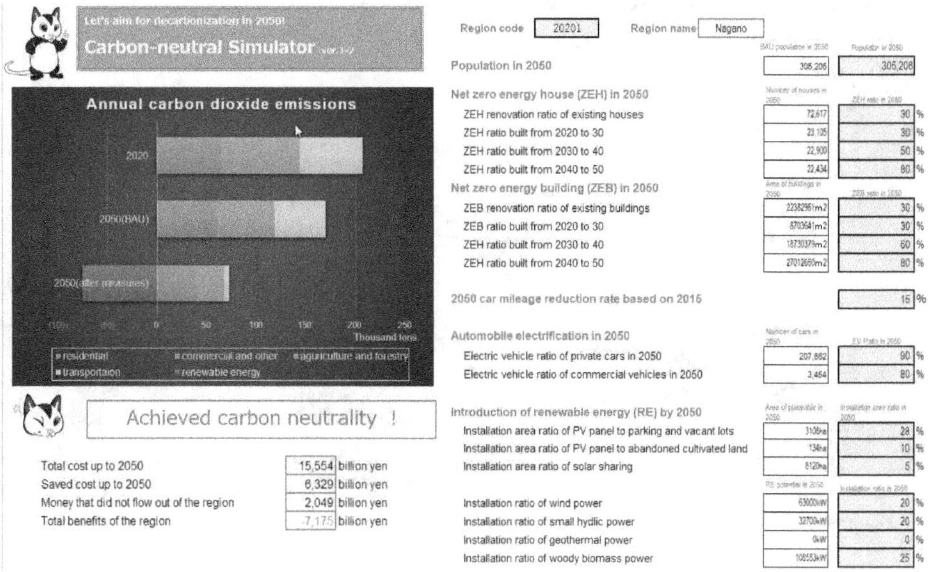

Figure 3. Input/Output Screen of the School Version of CNS

Educational practice using CNS and the educational effects thereof.

Implementing education

An education program using the improved version of the CNS was implemented from June to July 2021 at two public middle schools (S Junior High School and T Junior High School) in different municipalities in Japan. The CNS education program was conducted face-to-face at S Junior High School but online at T Junior High School because of the impact of the COVID-19 pandemic.

At S Junior High School, we delivered lectures on climate change and decarbonization to 102 first- and second-year students. Thereafter, the students experienced the CNS. Throughout this period, we provided support for CNS operations to S Junior High School teachers.

At Junior High School T, we conducted a lecture on climate change and decarbonization with 120 second-year students from three classes (Figure 4). In the classroom, T Junior High School teachers provided easy-to-understand supplementary explanations of the content of our lecture. Thereafter, the students engaged with the CNS. During this period, T Junior High School teachers provided support for CNS operations.

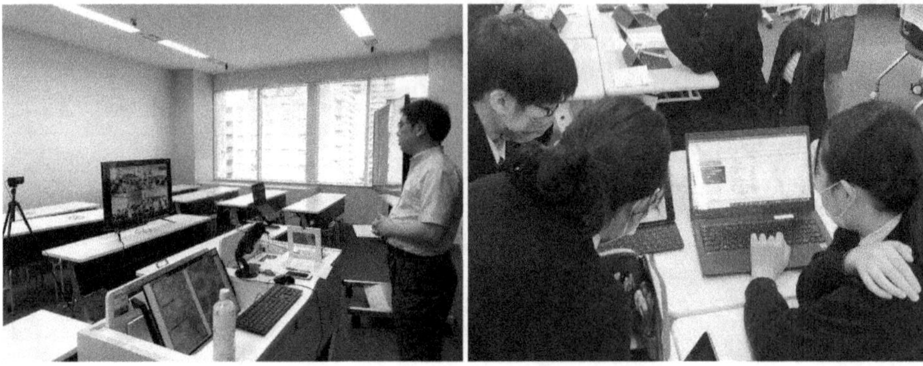

Figure 4. Scene of Online Class (T Junior High School)

Measurement of responses to CNS

Upon completion of the education program, we conducted a questionnaire survey to gather feedback from the participating students about CNS. The survey items assessed (1) operability (ease of use, screen visibility), (2) level of understanding, and (3) willingness to learn. The survey included six statements: "The CNS was easy to understand," "The CNS was easy to use," "The CNS data entry screen was easy to see," "I enjoyed learning about decarbonization," "I understand how decarbonization can be achieved," and "I understand how much it will cost to achieve decarbonization." There were five ways to respond to these questions: "very much," "quite a lot," "neutral," "a little," and "not at all."

A questionnaire survey was administered to 102 first- and second-year students at S Junior High School and 120 second-year students at T Junior High School. The number of valid student responses was 99 for S Junior High School (first- and second-year students) and 120 for T Junior High School (second-year students). Figure 5 shows the results.

Overall, T Junior High School showed a greater evaluation of the CNS than S Junior High School. At S Junior High School, many students took all classes at once, but at T Junior High School, 120 students were divided into three classes, and each class received close teacher support. This is thought to have increased their evaluation in junior high school.

When asked whether "the CNS was easy to understand," 34.3% of S Junior High School students and 43.7% of T Junior High School students responded, "very much easy to understand." When the "quite easy to understand" response was included, 80% of students responded that the CNS explanation was "easy to understand."

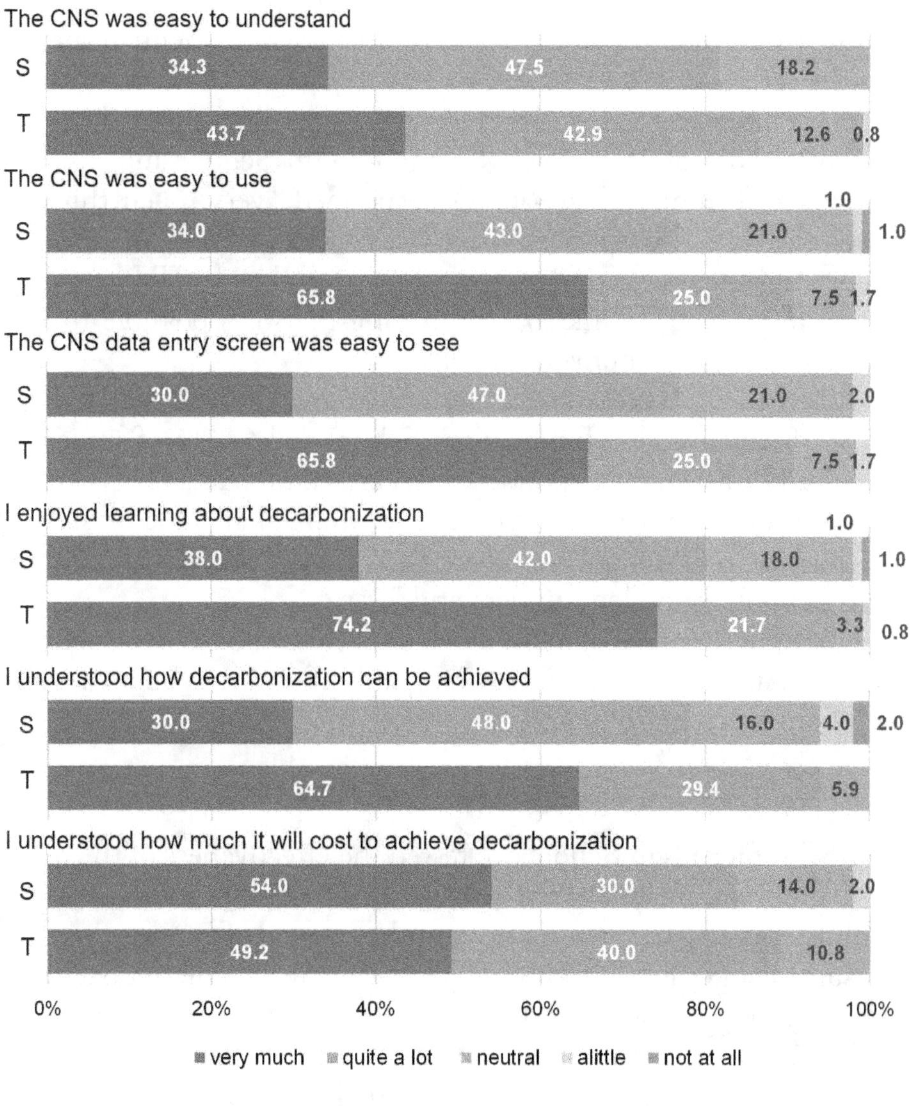

Figure 5. Evaluation of CNS by Students. (S Junior High School: n=99, T Junior High School: n=120)

When asked whether "the CNS was easy to use," 34.0% of S Junior High School students and 65.8% at T Junior High School responded, "very much easy to use." When the response "quite easy to use" was added, more than 80 to 90% of students at both middle schools responded, "easy to use," confirming that the school version of CNS was sufficiently easy to use even for junior high school students. When asked whether "the CNS data entry screen was easy to see," 30.0% of S Junior High School students and 65.8% of students at T Junior High School responded, "very much easy to see." When the "quite easy to see" response was included, about 80% of students responded that the screen was "easy to see."

When asked whether "[they] enjoyed learning about decarbonization," 38.0% of students at S Junior High School and 74.2% of students at T Junior High School responded, "very much think so." When the response "quite think so" was added, 80% of students at S Junior High School and more than 90% of students at T Junior High School responded, "think so," confirming that it is possible to enjoy learning about decarbonization with CNS.

When asked, "I understand how decarbonization can be achieved," 30.0% of students at S Junior High School and 64.7% of students at T Junior High School responded, "very much understand." When the response "quite understand" was added, more than 70% of students at S Junior High School and more than 90% of students at T Junior High School responded, "understand."

From the above results, it is clear that the CNS was generally highly evaluated and could provide good information about regional decarbonization.

Verification of the Effectiveness of the Educational Program

At the S Junior High School, we tested the effectiveness of the overall CCE program, including the CNS experience. Questionnaires were administered to participating students before and after the program to measure changes in their knowledge of climate change (global warming), decarbonization, the situation in their home city S, changes in their attitudes toward climate change (global warming) and decarbonization, and their contributions to the local community.

The response options were "very much," "quite a lot," "neutral," "a little," and "not at all" on a 5-point scale, with "knowledge" being comprehension and "attitude" being the change in awareness.

The survey was administered to 102 first- and second-year students at S Junior High School, with 100 valid responses from the pre-survey and 99 responses from the post-survey. Figure 6 illustrates the results of this analysis.

Overall, knowledge and attitudes improved after the implementation of the educational program. Notably, there was a significant enhancement in knowledge of decarbonization, climate change, and ways to mitigate carbon dioxide emissions. Regarding attitudes, there was a substantial increase in students' awareness and desire to improve the future of their city. In addition, students' engagement in actions to address global warming (climate change), which was already at a high level prior to the program, witnessed an almost 10-point improvement following program implementation.

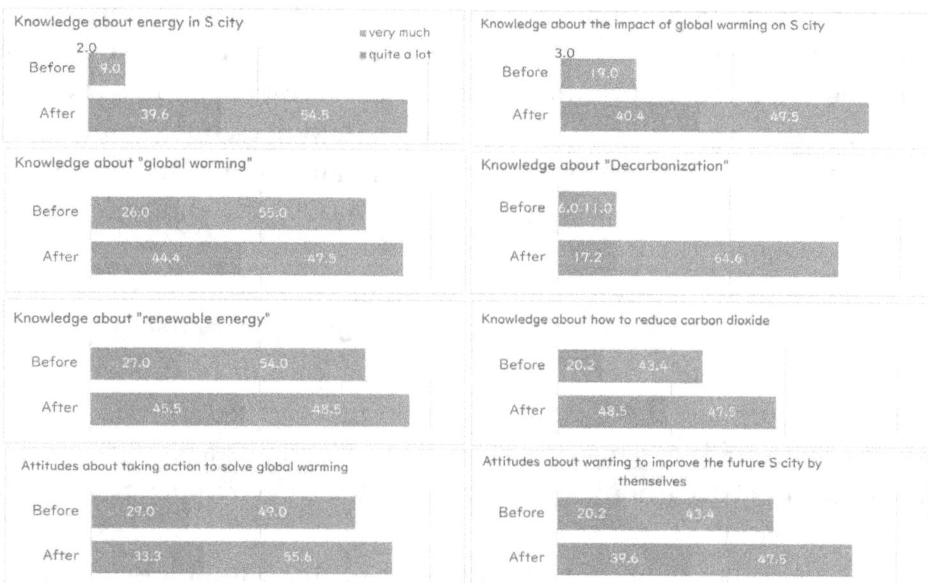

Figure 6. Changes in Knowledge and Attitudes Before and After the Educational Program in S Junior High School (Before: n=100, After: n=99)

The most striking feature of the survey results is the improvement in knowledge and attitude toward S City, where they live. The level of understanding of items such as "the energy situation in S City" and "the impact of global warming on S City" was very low before the implementation of the educational program but was confirmed to have improved significantly after the educational program, which included the CNS experience. Furthermore, awareness of improving S City's future has increased drastically. The CNS and this educational program are based on community-specific data; therefore, the students could understand how their city would be affected by climate change and the energy situation in S City.

Conclusion

In this study, we developed a school version of the CNS and used it to develop and implement a CCE program that raises students' awareness of a decarbonized society.

It is difficult for middle school students to understand climate change and decarbonization. Moreover, it is not possible to immediately predict the future in 2050. Therefore, in this study, we developed an educational program using the CNS, a tool that can easily simulate measures for each municipality to achieve carbon neutrality in the household, transportation, and business sectors by 2050. However, in the initial version, it was difficult for middle school students to understand the software operation and simulation results. Therefore, we developed a school version of the CNS that is friendly to middle school students.

The students received the CCE program well using the school version of the CNS, and their knowledge and attitudes improved. By simulating the experience of achieving carbon neutrality in the regions where they lived, students could perceive decarbonization as their responsibility. Using the CNS allowed students to enjoy learning about regional decarbonization, which fostered their understanding of this technology.

On the other hand, it is not possible to make middle school students understand climate change and decarbonization with "tools" such as CNS alone. In parallel with the development of the CNS, it is possible to obtain sufficient effects by improving CCE programs according to the developmental stages of middle school students and by implementing educational practices in collaboration with teachers.

Acknowledgments

This research was supported by the Environment Research and Technology Development Fund (JPMEERF20192001) of the Environmental Restoration and Conservation Agency Provided by the Ministry of Environment of Japan and JST COI-NEXT program Grant Number JPMJPF2003.

References

Brumann, S., Ohl, U. and Schulz, J. (2022). Inquiry-Based Learning on Climate Change in Upper Secondary Education: A Design-Based Approach. *Sustainability*, 14(6), 3544.https://doi.org/10.3390/su14063544.

Cordero, E.C., Todd, A.M. and Abellera, D. (2008). Climate Change Education and the Ecological Footprint. *Bulletin of the American Meteorological Society*, 89 (6), 865–872. https://doi.org/10.1175/2007BAMS2432.1.

Galeote, D.F., Rajanen, M., Rajanen, D., Legaki, N., Langley, D.J. and Hamari, J. (2021). Gamification for Climate Change Engagement: Review of Corpus and Future Agenda. *Environmental Research Letters*, 16(6), 063004. https://doi:10.1088/1748-9326/abec05.

Hallar, A. G., McCubbin, I.B. and Wright, J.M. (2011). CHANGE: A Place-based Curriculum for Understanding Climate Change at Storm Peak Laboratory, Colorado. *Bulletin of the American Meteorological Society*, 92 (7): 909–918. https://doi:10.1175/2011BAMS3026.1.

IPCC (Intergovernmental Panel on Climate Change). (2018). *Special Report on Global Warming of 1.5°C.* https://www.ipcc.ch/sr15/

IPCC (Intergovernmental Panel on Climate Change). (2021). *Climate Change 2021: The Physical Science Basis.* https://www.ipcc.ch/report/ar6/wg1/

IPCC (Intergovernmental Panel on Climate Change). (2022). *Climate Change 2022: Impacts, Adaptation and Vulnerability.* https://www.ipcc.ch/report/sixth-assessment-report-working-group-ii/

JMA (Japan Meteorological Agency). (2020). *Climate Change in Japan 2020.* https://www.data.jma.go.jp/cpdinfo/ccj/2020/pdf/cc2020_gaiyo_en.pdf

Klosterman, M.L. and Sadler T.D. (2010). Multi-level Assessment of Scientific Content Knowledge Gains Associated with Socioscientific Issues-based Instruction. *International Journal of Science Education*, 32 (8), 1017–1043. https//doi:10.1080/09500690902894512.

MEXT (Ministry of Education, Culture, Sports, Science and Technology, Japan). (2020). *About GIGA School Program.* https://www.mext.go.jp/en/content/20200716-mxt_kokusai-000005414_04.pdf

MOE (Ministry of the Environment, Japan). (2021). *2020 Public Opinion Polls on Climate Change.* https://survey.gov-online.go.jp/r02/r02-kikohendo/index.html

MOE (Ministry of the Environment, Japan). (2022). *2050 Zero Carbon Cities in Japan.* https://www.env.go.jp/en/earth/cc/2050_zero_carbon_cities_in_japan.html

MOE (Ministry of the Environment, Japan). (n.d.). a *Climate change adaptation information platform (A-Plat).* https://adaptation-platform.nies.go.jp/en/

MOE (Ministry of the Environment, Japan). (n.d.). b *Renewable Energy Potential System (REPOS).* https://www.renewable-energy-potential.env.go.jp/RenewableEnergy/

Monroe, M.C., Plate, R.R., Oxarart, A., Bowers, A. and Chaves, W.A. (2017). Identifying effective climate change education strategies: a systematic review of the research. *Environmental Education Research*, 25(6), 791-819. https://doi.org/10.1080/13504622.2017.1360842

National Institute of Population and Social Security Research. (2018). *Regional Population Projections for Japan: 2015-2045.* https://www.ipss.go.jp/pp-shicyoson/e/shicyoson18/t-page.asp

Rooney-Varga, J.N., Kapmeier, F., Sterman, J.D., Jones, A.P., Putko, M. and Rath, K. (2020). The Climate Action Simulation. *Simulation & Gaming*, 51(2), 114-140. https://doi.org/10.1177/10468781198906

Sterman, J., Franck, T., Fiddaman, T., Jones, A., McCauley, S., Rice, P., Sawin, E., Siegel, L. and Rooney-Varga, J.N. (2014). WORLD CLIMATE: A Role-Play Simulation of Climate Negotiations. *Simulation & Gaming*, 46(3-4), 348-382. https://doi.org/10.1177/104687811351493

UN FCCC (UN Framework Convention on Climate Change). (2015). *Adoption of the Paris agreement (FCCC/CP/2015/L.9/Rev.1).* http://unfccc.int/resource/docs/2015/cop21/eng/l09r01.pdf

Chapter 11

Forest kindergartens in the Czech Republic

Kateřina Jančaříková, Magdaléna Kapuciánová, and Karolína Kapuciánová

Abstract

Our paper outlines the movement in support of forest kindergartens in the Czech Republic, which was started by Emilie Strejčková. We owe a lot to Emilie Strejčková for her friendship, wisdom, and kindness and for the fact that thanks to her, we got to know each other, started working together, and even became friends.

In the last 20 years, we have seen incredible progress. In 2003, when Emilie Strejčková started promoting the need to establish forest kindergartens, there were none in the Czech Republic. In the present day (2023), there are about 200 of them in the Czech Republic, attended by over 3300 children.

We are lucky to have been able to personally experience this turning point in the history of pedagogy in the Czech Republic. We would like to report on this amazing change abroad.

Our paper is based not only on the usual scientific approaches but also on the life experience of Magdaléna Kapuciánová and Karolína Kapuciánová, the director of and a teacher from, respectively, one of the first forest kindergartens in the Czech Republic, along with the university lecturer Kateřina Jančaříková. Together, we have followed the movement out of deep interest.

Theoretical background

About the Czech Republic

The Czech Republic (also referred to as Czechia) is a landlocked country in Central Europe. Until 1990, the Czech Republic and Slovakia were part of the Eastern Bloc as the Czechoslovak Socialist Republic. Shortly after the fall of the communist regime, the Czech and Slovak republics parted ways peacefully (as of 1 January 1993). To understand Czech education, it is essential to remember that the Czech Republic was first part of the Habsburg monarchy and then, up until 1918, the Austro-Hungarian empire, so the first "Czech" school laws were associated with Empress Maria Theresa and her son Joseph. It is also worth noting that between 1938–1945, the Czech Republic was occupied by Nazi Germany.

The Czech Republic has a hilly landscape that covers an area of 78,871 square kilometres. The mixed deciduous forests (predominately of oak and beech), native to the landscape, have been replaced by spruce monocultures in most of the area. The population is over 10.5 million.

History of preschool childcare in the Czech Republic

Traditionally, like everywhere else in the world, grandparents were involved in caring for preschool children. The first public institution („opatrovna") caring for preschool children in our territory was founded by Jan Vlastimír Svoboda in 1832 (Opravilová & Uhlířová, 2017). Since then, the purpose of preschools has shifted from simply looking after the children of working parents to systematic education.

The development of kindergartens is related to the trend of employed mothers that began in the 20th century and continued under the socialist regime. Children were placed in nurseries at a very young age. Many adults have unpleasant memories of the nursery routine. This reduced the number of nurseries (associated with the socialist regime) in the early 1990s. At that time, thanks to the work of the eminent Czech psychologist Zdeněk Matějček, it was also a trend for mothers to care for their children alone up to the age of 3 to 4.

Nowadays, there are approximately 5400 kindergartens in the Czech Republic. Most preschool children (over 94%) attend kindergarten. There are both state and private kindergartens in the Czech Republic today. A typical state-run kindergarten class has one teacher per 22 children (MŠMT, 2021b).

According to the Czech preschool curriculum (*Rámcový vzdělávací program předškolního vzdělávání*), all Czech children are supposed to be in contact with nature. The Czech curriculum uses the words "nature" or "natural" 20 times in 46 pages. Several of its goals are related to nature, e.g. visiting nature, learning about nature, the joy of nature, nature conservation, etc. (MŠMT, 2021a). However, the reality is very often unsatisfactory. By the beginning of the 21st century, the average time that preschool children in the Czech Republic were spending out of doors had dramatically decreased (many children were spending outdoors less than 2 hours a day). Even the Survey on the Proportion of time spent indoors and outdoors in kindergartens showed that some kindergartens go out for only 45 minutes. In the rain (or risk of rain), they do not go outside at all (Rubešová, 2010).

Forest kindergartens' establishment in the Czech Republic

In the Czech Republic, the forest kindergartens movement began to take shape shortly after 2000, thanks to Emilie Strejčková (1939–2009). Emilie Strejčková significantly influenced environmental education in the Czech Republic. Among her greatest contributions are the establishment of the first research team that dealt with children's relationship to nature (The Alienation from Nature, 2003-2005), the founding of the "Toulcův dvůr Environmental Education Centre" (1995) and the focus on preschool children. Until then, environmental education in our country was a topic for pupils aged 15 plus.

In 2003, Emilie Strejčková founded the *Semínko* (The Little Seed) kindergarten in Prague, which fulfils many of the ideas of the forest kindergarten. As soon as the kindergarten was established, the children's learning started in the outdoor environment. First, in the natural garden, among the flowers, herbs, and vegetable beds. From the beginning, however, the second place of education has been the natural surroundings of the environmental education centre, the fruit and forest trees, dry-loving and wet-loving plants, and pastures of domestic animals. However, the legislation of the time did not allow for a greater extent of a forest type of school. In 2009, the first forest kindergartens were established in the Czech Republic (still without legislative support). In 2010, *Lesníček* (The Little Forester) was founded as the first true forest kindergarten (in accordance with Czech legislation).

The process of legalization of the forest kindergarten school program was long and arduous. The first success was that the Ministry of Education, Youth and Sports of the Czech Republic allowed the experimental validation of *Lesníček* (2010). The aim of the experimental verification was to determine whether and under what conditions it is possible to integrate a forest kindergarten into the operation of a traditional kindergarten.

The pilot project *Lesníček* integrated forest pedagogy into the preschool education curriculum. A pedagogical concept of forest kindergartens and a comprehensive methodology (education, hygiene, accident prevention) were developed and verified. A comparison of morbidity, school readiness, and other indicators of the children from *Lesníček* class with other children from mainstream kindergartens was also done (MŠMT, 2010).

The results of the pilot survey were published in interim reports. In these documents, the founders of other forest kindergartens in the Czech Republic found useful information for the multiplication of forest kindergartens in the Czech Republic (NÚV, 2010).

The final evaluation of the pilot validation confirmed the assumption that forest kindergartens can be an alternative to mainstream preschool education (2012). As a result, an amendment of the legislation was initiated, which allows the establishment of forest kindergartens entirely independently of the background of a regular kindergarten. It has been a long but successful process.

In December 2010, teachers and people who wanted the development of forest kindergartens in the Czech Republic jointly founded The Forest Kindergarten Association (*Asociace lesních mateřských* škol). Its aim is a) to take care of good conditions of preschool education in nature and to facilitate the quality and sustainability of forest clubs and kindergartens in the Czech Republic, b) to link Czech forest kindergartens to foreign countries (ALMŠ, 2018b, *Asociace lesních mateřských* ško*l* website—*www.lesnims.cz*). Thanks to this association, it was possible to avert the threat in 2014. That year, a new education law was being prepared in the Czech Republic that did not provide for this form of school attendance, and forest kindergartens were therefore threatened with extinction.

Since the academic year 2017/2018, forest kindergartens in the Czech Republic have become recognised by the legal system as full-fledged preschool education.

Czech Forest kindergartens' background

Semínko and other forest kindergartens in the Czech Republic follow Ella Flatau´s ideas. She established the first forest kindergartens in their present form in mid-20th century Denmark (Kane & Kane, 2011; MacQuarrie, Nugent, & Warden, 2015). Interestingly, the founders neglected the Czech tradition of outdoor learning, which was promoted by the representatives of eubiotics (Štorch, Růžička, etc.) after Czechoslovakia gained independence in 1918. This is probably because, as Czech Scout representatives, they had been erased from pedagogy textbooks. Both Nazis and communists banned scouting in our country (Jančaříková, 2019).

Today, forest schools can be found in dozens of countries worldwide. Their programme is as diverse and unique as their locations, students, and teachers, e.g., in Denmark, Sweden, Germany, the United States, the United Kingdom, Australia, New Zealand, Canada, China, Japan and Italy (Lubbock, 2020); Slovakia, Poland, and Hungary (ALMŠ, 2020); and Lithuania (ALMŠ, 2018a).

About 200 preschool groups are now involved in the forest kindergarten movement (Valkounová, 2022). Many of them (160) are members of the Czech Forest Kindergarten Association. In 2022, 888 people worked in these 160 preschools associated with the Forest Kindergarten Association, and they cared for 3367 children (ALMŠ, 2022).

For some time, forest kindergartens have also constituted a topic of bachelor and master theses (for example, Kapuciánová, 2010; Bendíková, 2022). Moreover, three articles by Czech authors have been published in Web of Science (Michek et al., 2015; Domljan, Šimek, Iliev, 2020; Cibulková, 2021).

Czech Forest Education movement

The movement in support of forest primary schools is a continuation of the forest kindergartens movement in the Czech Republic from a few years ago. The parents of Forest Kindergarten children have been looking for ways to continue outdoor education at the primary school level. The establishment of *forest schools* is a trend nowadays. Unfortunately, Czech legislation does not allow community schools. The parents of these children, therefore register them as home-schooled.

Home-schooling has been possible in the Czech Republic since 1997 (Kostelecká at all, 2016).

The parents and educators from these alternatives (home-schooling, forest kindergarten, outdoor education, and community schools) significantly influence traditional Czech education. The Lessons in Grass movement's slogan, *"Go out!"* from 2017, has had a powerful influence (TEREZA, 2022). Its goal is to promote outdoor learning and support the transformation of school gardens into natural ones.

We will focus on the "garden, forest and outdoor primary school movement in the Czech Republic" in our future research.

Methods

We use qualitative research methods. Our approach is based on the grounded theory method (Strauss & Corbin, 1998). Our goal was to use the description of *Lesníček* as an output of scientific work. In this case, we consider the description very important as abroad; there is not enough awareness of forest kindergartens in the Czech Republic.

Our description is based on long-term observation (more than 10 years). Two authors (Magdalena Kapuciánová and Karolína Kapuciánová) have been monitoring the situation inside the kindergarten (the first as the director and the second as a teacher in the kindergarten) and the third (Kateřina Jančaříková) as an outside expert. We have been writing reports, keeping diaries, taking photos, and shooting videos throughout this period. We decided to present a day and a year in *Lesníček*.

In the autumn of the academic year 2021/2022, a SWOT analysis was conducted, i.e., external and internal strengths, weaknesses, opportunities, and threats (Miranda et al., 2022). The respondents of the analysis were eight teachers, 20 parents (mothers and fathers) and 10 children (who are all currently attending or were attending *Lesníček* in the past). We followed up with interviews with teachers and parents, and some of the findings from the SWOT analysis were discussed (spring 2022).

Results

Pre-primary education in Lesníček class: general information

Lesníček (The Little Forester) is one of the founders of the Association of Forest Kindergartens. It is one of the three classes in Kindergarten *Semínko* (The Little Seed) in Toulcův dvůr, the environmental education centre in Prague. The forest class *Lesníček* was founded in 2010 as a pilot testing project of the Czech Ministry of Education, Youth and Sports. In 2012, this project won a UNESCO award as a *Project for Sustainable Development*Fifteen children are registered in a mixed-age class (boys and girls aged 3-7). Three teachers regularly work with the children. Two of them are always present. The kindergarten makes an effort to have a mix of female and male teachers.

The educational program is carried out outdoors, regardless of weather, in the natural area of the Environmental Education Centre Toulcův dvůr. In addition, the natural surroundings within 5 km are used. The teachers and children sometimes go on trips to more distant parts of the city: zoos, botanical gardens, galleries, museums, or libraries.

In case of bad weather, there is a mobile trailer for year-round use. In addition, a classroom is available in case of extreme conditions. It is furnished with natural materials, has a thick carpet of sheep's wool on the floor and is painted in the spirit of the stories of the Little Prince by Antoine de Saint-Exupéry. The children use this for about 2/12 of the year, not so much because of the cold but because of the drying space it provides.

In nature, mental but also manual and physical resilience is strengthened. In addition, the children's mental, emotional, and social development is positively affected. The most important thing is that being outdoors boosts immunity and helps to prevent illness. It also serves to prevent injuries, addictions, allergies, etc.

Playing outdoors promotes creativity, communication skills, and planning, but also equally important elements such as self-confidence, perseverance, and the ability to concentrate. The environment inspires the children to be creative and skilful and boosts their imagination. In nature, it is possible to experience things more acutely and with all the senses.

In school education, natural building materials and environmental creation are used.

It is used as a worktable for production and repairs. The children use real tools.

They can prepare and cook their own food. They cook it on the fire and after they eat it.

Meals

Meals and eating together are considered very important because the ritual of eating together is also disappearing from family life. The children have their snacks outside as a picnic. They have lunch in the kindergarten dining room. They also prepare ingredients for their own cooking or baking in the canteen if they are not cooking over the fire.

Hygiene

Natural facilities include a separate composting toilet and a sink with running hot water from a canister for personal hygiene. Organic soap is used for washing. Children dry their hands with machine-washable cloth towels. Outside the facility, children relieve themselves in the open air. There is a flush toilet with a sink in the bricked-in back room in the Little Prince classroom. The kindergarten building has a flush toilet and a bathroom with a shower.

Close contact with parents

Every morning and afternoon, parents have the opportunity to talk to the teachers. There are regular monthly meetings between parents, children, and teachers. Thanks to these festivals, the teachers have closer relationships with the parents, who are often present at the events.

Close contact with animals

The children in *Lesníček* are in regular and direct contact with animals in the kindergarten (a bearded agama, stick insects, fish, or the therapy dog: a border collie called Dora) or on the neighbouring farm (rabbits, chickens, sheep, goats, horses, cows etc.). The children are also in contact with wild animals—birds that they feed at bird feeders, aquatic and terrestrial arthropods and amphibians, and snails that they catch and observe.

Social area

Intergenerational meetings are implemented in *Lesníček*. The teachers from the school invite the children's grandparents out for a day. On this day, the children look after their grandparents. They prepare snacks for them, dance with them, sing for them, draw their portraits, and play with them (MŠ Semínko, 2019). Accompanied by the teachers, the children regularly visit nursing homes for the elderly to sing, dance, and make things together (Kubejová, 2022).

How legislation differs for normal classes and *Lesníček* class

The legislative requirements for the forest class are the same as those for the normal class. In addition, the forest class has a natural background, but this is not a legislative requirement. The advantage of this arrangement is that children can have space inside if needed. If not needed, the indoor space can be used for workshops, special groups, or further education.

A typical day in Lesníček

Beginning

In the morning, the teachers arrive at work, pack the day's provisions and supplies in their backpacks, prepare tea for themselves and some extra for the children, dress in outdoor clothes, and head out into the natural environment to wait for the children.

Then, the parents bring the children and greet the teachers with handshakes. There is space for very brief organizational conversations with the teachers (a child's exceptional condition an important event in the family). The children play together freely, look at books, or play a game with the other teacher.

Morning circle

The structured programme begins with the morning circle. It is a time to make the children present, "Now we are together at the kindergarten". In the circle, the children learn to feel their community, count each other, and find out who is there and who is missing. In the morning circle, topics related to the time of year and the situation around the children are also addressed, and gradually, the motivation for the planned activities of the day is approached.

The afternoon guided program

Structured education takes the form of an invitation. A transfer (on foot, on bicycles) to a suitable location usually precedes it. Fine and gross motor skills, all competences and pre-literacy are developed naturally (children collect firewood, cook, make things, draw, make music, and climb slopes and trees). The programme ends with an evaluation, a summary of what the children have learned.

Lunch

Afterwards, they go to the school dining room (approx. 250 m). After lunch, the children return to the forest kindergarten base and rest in their sleeping bags on sleeping mats under the treetops or in the mobile house. Their teacher reads them a story. Those who do not fall asleep look at a book or get dressed and play freely.

Free play

Spontaneous free play is important in a child's life. Play is the child's means of learning. In free play, the child has enough space to experiment and evaluate the results independently; in group play, the child learns in different roles (dominant, submissive, etc.). New ideas are generated during free play, and creativity and imagination are developed. The teacher enters the free play minimally and stands aside in the role of an observer but also a protector.

The children have space for free play in the morning before they all meet, after snacking before they all finish their food, and at the end of the day before their parents pick them up. Free play is also incorporated during the morning between the individual activities controlled by the teachers or instead of it (on a trip, etc.).

Handing over the children to their parents

The parents collect their children at 3 pm. Again, there is space for short conversations with the teachers.

Teacher-related work

After the teachers have handed the children over, they write a report to the parents on the e-board about what they experienced that day. This way, parents know what happened and can ask the children guiding questions to help them tell their story. They also post photos. This e-board is closed to the public, and parents can see it after entering their password. Some parents make a diary of these messages and photos for their children. Sometimes, teachers have consultation sessions with parents; sometimes, they have joint meetings with other teachers in the school about joint events. Finally, the teachers prepare the supplies for the next day.

A typical day is described in more detail in the school's educational programme called "The whole year's harvest depends on the spring seeding" (MŠ Semínko, 2021).

Academic year 2021/2022 at the Forest Kindergarten Lesníček

September—The children got to know their teachers, each other, and their environment and set rules together.

Autumn workshops—children's and parents' joint projects, tasting of healthy homemade pastries from the school canteen and from home.Dental students came to speak to the children about dental hygiene.

The children and teachers from the whole kindergarten took part in the Clean up the Czech Republic event—cleaning up litter in nature.

Michael's Feast was held—the teachers and children bought vegetables, cleaned them, cut them and cooked soup. Refreshed by the soup, the children went into the forest on their own or with the teacher's accompaniment to find the treasure. The treasure was a red apple.

October—Gifts of Autumn

The children harvested what they had grown in the vegetable patch—the last tomatoes, peppers, zucchini, and pumpkins. They picked fallen apples and nuts and collected chestnuts for animals.

The children dyed silk with natural dye from boiled walnut leaves and made elves.

Autumn work session—joint improvement of the kindergarten premises by the parents, children, and the school team, sitting around the fire singing and roasting sausages and playing musical instruments together.

November—St. Martin Arrives on His White Horse

Preparations for the St. Martin festival, children made lanterns, listened to the story of St. Martin, and participated in a school play. Then, a real horse arrived with St. Martin riding on it, and it helped the beggar, sharing St. Martin cakes with each other, parents, and teachers.

The children pressed fallen leaves, stamped them, made leaf crowns and kites, and compared sizes and types of leaves.

At the open day, the teachers welcomed prospective parents and children.

The children cooked porridge over the fire, roasted "trdelník" (cinnamon pancake) and prepared sauerkraut.

December—Advent, Christmas Time

The children baked Christmas cookies from organic food bought in a packaging-free shop and made yoghurt. In the pottery workshop, they made mugs for their parents.

Christmas party—the children performed in front of the church, the teachers played the violin and flute, and everyone sang. Singing and chatting, exchanging Christmas recipes between parents, and Christmas cookie tasting.

January—Birds

The children organized a Three-Magi parade for the neighbours.

No leaves were on the trees so the children could observe and recognize the birds. The children recognized different species, observed them through binoculars, recognized their bird songs, etc.

The Lunar Year was celebrated as one family came from Vietnam. To celebrate the Lunar New Year, the parents prepared a talk with Vietnamese fruits and specialties tasting.

February—Mardi Gras

For the parents' evening, which was held outside in the garden due to epidemic measures, the children and their cooks prepared soup—meat and vegetable broth—for the parents. The children cleaned and chopped all the vegetables.

The main Mardi Gras week topic was "Grandmothers and Grandfathers". Through various games and activities, the children's competences of intergenerational respect, communication, help, and mutual enrichment were promoted. During the parade, together with the teachers, the children sang to the workers from neighbouring organizations and covered their faces with soot for good luck.

March—Healthy Living

Lesníček started pre-growing their own vegetables, and the children observed the baby chicks born on the neighbouring farm. They went on long hikes. They also visited the library because March is the month of books. Together, they made a wooden doll, the protector of the place.

Medical students came to the preschool with a programme called "Teddy Bear Hospital". The children brought soft toys and pretended that they were the patients that they, as doctors, treated.

April—Spring is Coming

Easter workshops, customs and traditions—joint children-and-parents projects, tasting of healthy homemade pastries from parents and spreads from cooks. They also sowed microgreens (peas, horseradish, and watercress).

Kateřina Jančaříková came to the kindergarten to present a programme about snakes. With the help of this model organism, she introduced snakes and their life to the children. With models, she practised mathematical thinking with the children.

Spring work session—a spring clean, establishing flower beds, decorating the school premises by the parents, children and team, sitting around the campfire with singing and sausage-roasting, playing musical instruments together.

May -The Time of Love

In May, we celebrated Mother's Day. The teachers prepared workshops for the fathers and their children and then celebrated in *Lesníček*. Together, they made a wooden heart in a vice for the mummies using tools such as a saw, rasp, and file. As another gift, the mothers also got a free afternoon without children.

The Grandparents' Café was a great celebration—healthy food prepared by the children was tasted, and the children sang and danced with their grandparents.

Moreover, the children went to the nursing home, where they played percussions along with the elderly.

June—Water, Preparation for Summer Holiday Adventures

The children fished for animals in the wetland, sailed down the nearby Botič stream, participated in the local kindergarten Olympics, and the oldest went on a weeklong expedition together.

The year closed with a "farewell to preschool kids"—a children's play, the giving of gifts and small messages for the academic year, the release of butterflies that hatched in the kindergarten and a sleepover in the garden.

The academic year is described in more detail in the *Semínko* Annual Report (MŠ Semínko, 2022).

SWOT Analysis Outputs

The main **strengths** of *Lesníček* are as follows: a small group of children, two teachers throughout the lessons, and an environment allowing contact with nature, specifically Toulcův dvůr area, the garden, the animals (for more, see Table 1).

Table 1. Strengths of Lesníček kindergarten.

Parents	Teachers	Children
Small group of children	Small group of children	Friends
Two teachers present at all times	Facilities of the kindergarten (food, shower, toilet)	Teachers
Healthy food	Better immunity of the children	Swing Mobile house
Close contact with nature	Close contact with nature, nice environment	Garden, animals

The parents, teachers, and children view the weaknesses very differently. While the parents worry about the cold, teachers and children do not. The teachers see that visitors to the *Toulcův dvůr* area can walk right up to the kindergarten premises as a weakness and express concerns about the risk of vandalism. The children see a problem in the fact that the area is muddy in the winter (for more, see Table 2).

Table 2. Weaknesses of Lesníček kindergarten.

Parents	Teachers	Children
Short operating time		Low swings

Fear of cold	Strangers can destroy the facilities	It is more mud than grass in winter

Some previously mentioned elements are repeated in the **opportunities** category (relationship to nature, fear of vandalism, the garden) and some solutions are proposed (to build a fence). An expansion of the capacity is desired so that more children (younger siblings) can attend the forest kindergarten. As a contradiction to the idea of a forest kindergarten, it sounds like the wish of one of the children to "build a house" (for more, see Table 3).

Table 3. Opportunities of Lesníček kindergarten.

Parents	Teachers	Children
Graduates can have a better relationship with nature	Create more forest classes	Little brother join
Children will think about conservation	Build a fence to prevent unwelcome disturbance	More flowers
		Bigger garden
		Build a house

The fewest items appeared in the **threats** category. Parents mention fears of bad weather, legislative changes (which would ban forest nurseries) and lack of space (as do teachers). Children express fears of being alone and of being threatened by fantastic creatures and thieves (for more, see Table 4).

Table 4. Threats to Lesníček kindergarten.

Parents	Teachers	Children
Lack of space	Threat of lack of natural places to set up forest kindergartens in	A devil might come A monster might come A dragon might come

Legislative can be changed		A thief might come
Bad weather		Being alone there

Discussion

Similar to foreign countries, the trend of outdoor learning has been developing in the Czech Republic in recent decades. Parents, teachers, and professionals dissatisfied with public education have pushed for this alternative. They intended for children to spend more time in a natural outdoor environment that allows them to experience nature and builds their immunity, resilience, and self-control. Another advantage of education in forest schools is the smaller number of children in a class and the constant presence of two teachers. Forest kindergartens in the Czech Republic are age-heterogeneous, which helps newcomers to adapt. The movement has come a long way since the opening of the first forest school in the Czech Republic in 2010 (Kapuciánová, 2010), and the pool of forest kindergartens that can support each other has expanded many times over.

The Comparison of a Forest Kindergarten and a Regular Kindergarten

Both regular and forest kindergartens care for children whose parents need childcare during their employment; they also seek a place for their children to socialize, make friends and learn to get along with their peers in small mixed-age groups. A traditional Czech state kindergarten sometimes has as many as 28 children per teacher. The teachers themselves prefer to work in homogeneous groups (Syslová et al., 2021), although from some points of view, it has many disadvantages (Koťátková, 2014). The children spend most of their time indoors (Rubešová, 2010).

Parents who are interested in educating their children in year-round daily contact with nature with stress on sustainable development enrol their children in the Forest Kindergarten. Education and nature are equally important to them. The goal of being in nature is to educate children and teach them not only colours and numbers but also the awareness of the primary conditions of life on Earth through actual experience with them (Vošáhlíková, 2010).

The COVID-19 period, in Comparison to the Pperation of *Lesníček*, Regular Glasses in the *Semínko* Kindergarten, and the Common Kindergarten

At the time of Covid 19, the children in *Lesníček* were taught as before—outdoors in nature. In addition, the stay in nature was extended to include meals. The school canteen was also moved outdoors to the garden pavilion. Except for when all kindergartens in the country had to be closed (11 days), the forest school *Lesníček* was running as usual. On the days when it had to be closed, education was conducted remotely. The children were at home with their parents. The teachers sent education tips to the parents.

The garden classes of the Kindergarten *Semínko*, i.e. the regular classes, were at that time similar to the operation of a forest kindergarten. The children's education also took place outdoors during meal and resting times. In contrast, education in common kindergartens was very limited. The regular kindergartens were closed for two months. Later, after the reopening, much of the education in the common kindergartens also moved outdoors (MŠMT, 2021c), following the example of the forest schools. Hopefully, this trend will continue to grow, and more time will be spent outside in common kindergartens, as is the case in Norway, where the use of different outdoor environments is a common everyday teaching practice regardless of weather (Melhuus, 2012).

The question is whether some families would choose this alternative if the children in the traditional state kindergartens went out for longer periods if the number of children was smaller, and if the number of teachers increased. One child in the SWOT analysis wished to have *a house* on the Forest Kindergarten premises.

Daily Routine

The analysis of a day in *Lesníček* showed that the routine and rituals that are the same every day rule the day. This is not something new: it is the traditional approach in Czech kindergartens (Koťátková, 2014). Helle Hackmann, the founder of a forest kindergarten in Denmark and author of several books widely read by forest educators, also talks about the necessity of rituals (Hackmann, 2019). She puts her theory into five keys to a fulfilled life with children from birth to 7 years of age (she is also a Waldorf teacher, that is why it is 7 years).

SWOT Analysis

Although forest kindergartens are now a legally established part of the Czech education system, according to the SWOT analysis, many parents are concerned about potential changes in legislation and the loss of this alternative for their youngest children.

An interesting finding of the SWOT analysis is that some parents are worried about bad weather and cold, while teachers and children (i.e., those exposed to the weather) are not.

A very interesting finding is that the SWOT analysis did not include the fear of injury to children (falling from a tree, being injured by tools, being burned) or the lack of hygiene (no running water, outdoor toilet). These are the usual objections of people who do not support this alternative. As well as the most common issues addressed by inspections and approval authorities.

A key recommendation for the future identified by the SWOT analysis is the need for fencing to provide the kindergarten with more privacy. Common kindergartens in the Czech Republic are fenced. On the other hand, forest nursery areas usually are not. The fence provides security from unwelcome visitors (vandals, thieves—as mentioned in the SWOT analysis) and eliminates the children's undesirable departure from the premises. It is questionable whether the presence of a fence would reduce the children's fear of the arrival of a thief, monster, dragon, or devil expressed in the SWOT analysis. This topic was also discussed in the 2010 research entitled Forest Schools (Kapuciánová, 2010) in focus groups of the founders of forest schools, and freedom and safe space at the base camp were found to be the pillars of forest schools. The interesting thing is that unfenced space does not automatically mean no boundaries. In forest schools, borders are usually very strictly determined; however, they are marked psychologically rather than physically. For example, in *Lesníček*, there is a small raft made of wicker; in some schools, cotton strings are tied to the trees, representing boundaries. Bounded and unbounded space will be the research topic of the next thesis we commission.

Some other findings of the SWOT analysis also correspond to Kapuciánová's research (2010), e.g., among the advantages of forest class rank is the creation of a positive relationship with nature, a need for a greater number of teachers (more individual approach to children), better immunity. Also, concerns about insufficient hygiene, bad weather, or, paradoxically, higher morbidity among the children were mentioned; unfortunately, there was no division of the respondents into teachers and parents as is in this SWOT analysis, so a proper comparison cannot be made.

Need for Further Research

There are many topics that future research could be devoted to, for example, the morbidity of the children in forest kindergartens compared to common kindergartens, pro-environmental behaviour (or employment) among the children from forest kindergartens as adults, different forms of upbringing in forest kindergartens in comparison to common kindergartens etc.

Conclusion

Forest kindergartens are an alternative way of preschool education in the Czech Republic. Their popularity keeps growing. They are also increasingly attracting the attention of researchers. The forest primary school movement has recently followed the forest kindergarten movement, which is likely to develop at a similar pace in the coming years.

Resources

ALMŠ (12. 11. 2018a). *Litevská inspirace* [online, cit. 2023]. Avaible form: https://www.lesnims.cz/aktuality-na-uvodni-strance/litevska-inspirace.html

ALMŠ (20.12. 2018b). *Stanovy spolku Asociace lesních* MŠ, z. s. Praha.

ALMŠ (26. 2. 2020). *Země Visegrádu společně* řeší *lesní* školky [online, cit. 2023]. Available fromm: https://www.lesnims.cz/media/tiskove-zpravy/zeme-visegradu-spolecne-resi-lesni-skolky.html

ALMŠ (3. 6. 2022) *Výsledky jarního mapování: Kolik je v lesních* školkách *dětí a jaká je výše* školkovného*?*. [online, cit. 2023]. https://www.lesnims.cz/aktuality-na-uvodni-strance/vysledky-jarniho-mapovani-kolik-je-v-lesnich-skolkach-deti-a-jaka-je-vyse-skolkovneho.html

Bendíková, B. (2022). *Přehled výzkumných studií o lesních mateřských školách.* Bakalářská práce (Bc.). Univerzita Karlova, Pedagogická fakulta.

Bělinová, L., Jírová, M., Kořínková, H., Bartošová, E., Opravilová, E., Jůva, V. et al. (1963). *Rok* v *mateřské* škole. Praha: Státní pedagogické nakladatelství.

References

Cibulková, D., Francová, L., Růžička, I., Urbanová, D., & Suk, J. (2021). *Comparison of physical activity of children in a state kindergarten and a forest kindergarten in the Czech Republic.* Physical Activity Review, 2(9), 120-129.

Domljan, Danijela; Šimek, Milan; Dijanošić, Ivan; Iliev, Boris (2020). *Analysis of preconditions for design of sensory environment in Czech, Bularian, Macedonian and Croatian Kindergarten.* Sustainability of forest-based industries in the global economy, 28. – 20. 2020, Vinkovci, Croatia, WoodEMA, conference.

Hackmann, H. (2019). *Pět zlatých klíčů k naplněnému* životu *s dětmi od narození do sedmi let.* Praha: Malvern.

Jančaříková, K. (2019). *Didaktické přístupy k přírodovědnému vzdělávání předškolních dětí a mladších* žáků (*Second, expanded edition).* Univerzita Karlova, Pedagogická fakulta.

Kane, A., & Kane, J. (2011). *Wald kindergarten in Germany.* Green Teacher, (94), 16-19. Retrieved from http://search.proquest.com.ezproxy.library.uvic.ca/docview/904987533?acco untid=14846

Kapuciánová, M. (2010). *Lesní mateřské školy*. Diplomová práce (Mgr.)-Univerzita Karlova. Pedagogická fakulta, 2011.

Kostelecká, Y., Zimčík, L., Janaříková, K., & Jančařík, A. (2016). *Distance learning and the home schooling in the Czech Republic*. In: NOVOTNÁ, Jarmila – JANČAŘÍK, Antonín. Proceedings of the European Conference on e-Learning, ECEL. 1 vyd. Reading: Academic Conferences and Publishing International, 2016, s. 364-372.

Kubejová (2022). *Setkání na živo v rámci projektu Erasmus+ „Role and Opportunities of the Cultural Heritage in Intergenerational Learning". Chodováček* (6 - 7), 15.

Lubbock J. (2020). *A Brief History of Forest Schools Around The World*. Forest School Foundation.

MacQuarrie S., Nugent C., & Warden C. (2015). *Learning with nature and learning from others: Nature as setting and resource for early childhood education*. Journal of Adventure Education & Outdoor Learning, 15 (1), 1-23.

Melhuus, E. (2012). *Outdoor day-care centres–a culturalization of nature: how do children relate to nature as educational practice?*. European Early Childhood Education Research Journal, 20(3), 455-467.

Michek, S., Nováková, Z., & Menclová, L. (2015). *Advantages and Disadvantages of Forest Kindergarten in Czech Republic. Procedia, social and behavioral sciences*, *171*, 738-744.

Miranda, A. E., Rosadas, C., Assone, T., Pereira, G. F. M., Vallinoto, A. C., & Ishak, R. (2022). *Strengths, weaknesses, opportunities & threats (SWOT) analysis of the implementation of public health policies on HTLV-1 in Brazil*. Frontiers in medicine, 864.

MŠ Semínko (5. 6. 2019). *Novinky ze* škol – *Kavárna pro prarodiče* v *Semínku*. Avaible from Skutečně zdravá škola: https://www.skutecnezdravaskola.cz/novinky-ze-skol/skolni-blogy/kavarna-pro-prarodice-v-seminku

MŠ Semínko (2021). *Sklizeň celého roku záleží na jaře, kdy se seje: Školní vzdělávací program mateřské školy*. Praha.

MŠ Semínko (2022). *Annual Report of Semínko Kindergarten 2021/2022*. Praha. Awailable from Toulcův dvůr: https://www.toulcuvdvur.cz/files/f4efc87f345d9e9539a9f6be05da5501.pdf

MŠMT (6. 10. 2010). *Vyhlášení pokusného ověřování provozu integrované lesní mateřské* školy *Lesníček při Mateřské* škole *Semínko, o.p.s.* Praha. Avaible from: https://www.msmt.cz/file/10806?highlightWords=lesn%C3%AD+mate%C5%99sk%C3%A9+%C5%A1koly

MŠMT (2021a). *Rámcový vzdělávací program pro předškolní vzdělávání* . Praha:51 s. . Available from MŠMT: https://www.msmt.cz/file/56051/

MŠMT (2021b). *Vývojová ročenka* školství *2010/11 – 2020* [online, cit. 2023]. Avaible from MŠMT: https://www.msmt.cz/vzdelavani/skolstvi-v-cr/statistika-skolstvi/vyvojova-rocenka-skolstvi-2010-11-2020-21

MŠMT (2021c). *FAQ k první fázi návratu do škol* [online, cit. 2023]. Avaible form: https://www.msmt.cz/faq-k-prvni-fazi-navratu-do-skol

NÚV (2010). *Pokusné ověřování provozu integrované lesní mateřské* školy [online, cit. 2023]. Avaible from: https://archiv-nuv.npi.cz/t/pokusne-overovani-provozu-integrovane-lesni-materske-skoly-1.html

Opravilová, E., & Uhlířová, J. (2017). *Příběhy české mateřské školy: vývoj a proměny předškolní výchovy. 1. díl do roku 1948*. Praha: Univerzita Karlova, Pedagogická fakulta.

Rubešová, I. (2010). *Podíl času stráveného ve vnitřních a venkovních prostorách mateřské školy*. Bakalářská práce (Bc.). Univerzita Karlova, Pedagogická fakulta.

Strauss, A., & Corbin, J. (1998). *Basics of Qualitative Research: Techniques and Procedures for Developing Grounded Theory. 2nd edition*. London: SAGE Publications.

Syslová, Z., Nováková, E., & Najvarová, V. (2021). Vliv věkového uspořádání tříd na edukaci v mateřských školách. *Studia paedagogica (Brno)*, *26*(3), 109-130. https://doi.org/10.5817/SP2021-3-5

TEREZA (2022). *About us – Lessons in grass*. Available from Tereza: https://lessonsingrass.com/about-us/

Valkounová , T. (2022). Členská *základna*, Letní škola pro pedagogy LMŠ, 25.7.2022, Ekocentrum Loutí, ALMŠ, letní škola.

Vošáhlíková (2010). *Ekoškolky a lesní mateřské* školy. Praha: Ministerstvo životního prostředí.

Kateřina Jančaříková is a senior lecturer (docent) at the Teacher Training Faculty *of Charles University* and the *Faculty of Science of J.E. Purkyně University in* Ústí *nad Labem*, katerina.jancarikova@ujep.cz.

Magdaléna Kapuciánová is a teacher. Along with Emilie Strejčková, she founded *Semínko*, the kindergarten where nature teaches, 25 years ago. She is a Ph.D. student at the Teacher Training *Faculty of J.E. Purkyně University in* Ústí *nad Labem*, kapucianova.seminko@seznam.cz.

Karolína Kapuciánová is a teacher. She works at the *Semínko* kindergarten. She is a Ph.D. student at the Teacher Training *Faculty of J.E. Purkyně University in* Ústí *nad Labem*, kapagama@seznam.cz.

Images (Archive of Mateřská škola Semínko, o.p.s.)

Picture 1 Forest kindergartens attract a higher percentage of male teachers than common kindergartens in the Czech Republic (ALMŠ, 2022)

Picture 2 The children are preparing meals

Picture 3 Making tea from self- collected herbs over the fire

Picture 4 Airing the sleeping bags after the afternoon nap

Picture 5 Meeting the therapy dog

www.ingramcontent.com/pod-product-compliance
Lightning Source LLC
LaVergne TN
LVHW022206200926
848481LV00006B/1867